Springer Theses

Recognizing Outstanding Ph.D. Research

Aims and Scope

The series "Springer Theses" brings together a selection of the very best Ph.D. theses from around the world and across the physical sciences. Nominated and endorsed by two recognized specialists, each published volume has been selected for its scientific excellence and the high impact of its contents for the pertinent field of research. For greater accessibility to non-specialists, the published versions include an extended introduction, as well as a foreword by the student's supervisor explaining the special relevance of the work for the field. As a whole, the series will provide a valuable resource both for newcomers to the research fields described, and for other scientists seeking detailed background information on special questions. Finally, it provides an accredited documentation of the valuable contributions made by today's younger generation of scientists.

Theses are accepted into the series by invited nomination only and must fulfill all of the following criteria

- They must be written in good English.
- The topic should fall within the confines of Chemistry, Physics, Earth Sciences, Engineering and related interdisciplinary fields such as Materials, Nanoscience, Chemical Engineering, Complex Systems and Biophysics.
- The work reported in the thesis must represent a significant scientific advance.
- If the thesis includes previously published material, permission to reproduce this must be gained from the respective copyright holder.
- They must have been examined and passed during the 12 months prior to nomination.
- Each thesis should include a foreword by the supervisor outlining the significance of its content.
- The theses should have a clearly defined structure including an introduction accessible to scientists not expert in that particular field.

More information about this series at http://www.springer.com/series/8790

Robert Fickler

Quantum Entanglement of Complex Structures of Photons

Doctoral Thesis accepted by
the University of Vienna, Austria

Springer

Author
Dr. Robert Fickler
Vienna Center for Quantum Science and Technology
University of Vienna/IQOQI Vienna
Vienna
Austria

Supervisor
Prof. Anton Zeilinger
Vienna Center for Quantum Science and Technology
University of Vienna
Vienna
Austria

ISSN 2190-5053 ISSN 2190-5061 (electronic)
Springer Theses
ISBN 978-3-319-22230-1 ISBN 978-3-319-22231-8 (eBook)
DOI 10.1007/978-3-319-22231-8

Library of Congress Control Number: 2015946751

Springer Cham Heidelberg New York Dordrecht London

Printed on acid-free paper

Springer International Publishing AG Switzerland is part of Springer Science+Business Media (www.springer.com)

Publications Related to this Thesis

Quantum Entanglement of High Angular Momenta
Robert Fickler, Radek Lapkiewicz, William N. Plick, Mario Krenn, Christoph Schäff, Sven Ramelow and Anton Zeilinger, *Science* Vol.338, 640 (2012)

Real-Time Imaging of Quantum Entanglement
Robert Fickler, Mario Krenn, Radek Lapkiewicz, Sven Ramelow and Anton Zeilinger, *Scientific Reports* Vol.3 (2013)

Quantum Entanglement of Complex Photon Polarization Patterns in Vector Beams
Robert Fickler, Radek Lapkiewicz, Sven Ramelow, and Anton Zeilinger, *Physical Review A* Vol.89(6), 060301(R) (2013)

Entangled Singularity Patterns of Photons in Ince-Gauss Modes
Mario Krenn, Robert Fickler, Marcus Huber, Radek Lapkiewicz, William Plick, Sven Ramelow and Anton Zeilinger, *Physical Review A* Vol.87(1), 012326 (2013)

Quantum Orbital Angular Momentum of Elliptically Symmetric Light
William N Plick, Mario Krenn, Robert Fickler, Sven Ramelow and Anton Zeilinger, *Physical Review A* Vol.87(3), 033806 (2013)

Generation and Confirmation of a (100 × 100)-Dimensional Entangled Quantum System
Mario Krenn, Marcus Huber, Robert Fickler, Radek Lapkiewicz, Sven Ramelow and Anton Zeilinger, *Proceedings of the National Academy of Sciences*, 201402365 (2014)

Supervisor's Foreword

"*What is light?*" is a question which has fascinated humanity certainly since the earliest dates of written documents. But it is a fair guess that even the first humans, looking up at the night sky, wondered about the nature of light. Consequently, the study of the nature of light has accompanied the development of science in general and of physics in particular over all of its history. In the 20th century, light became *the* paradigmatic example in the debates on the conceptual challenges raised by the then new quantum mechanics. Einstein, as early as in 1905, when he proposed the existence of particles of light, realized the tension with the then prevalent wave picture. Many gedanken experiments, often using individual photons, the particles of light, served to underline the counterintuitive predictions of quantum physics for individual quanta. Real experiments were not possible in the 1920s and 1930s due to the rather limited technology at that time.

Technical progress, most importantly the invention of the laser, made experiments with individual photons possible. To date, numerous experiments have confirmed the dual wave and particle nature of light. Consequently, work with single photons is established routine in many laboratories around the world and has even entered student laboratory courses.

In his work, Robert Fickler combines two conceptually interesting features of light in novel and ingenious ways. On the one hand, there is entanglement, which is the correlation between distant measurement results without any possibility of a classical realistic explanation. Einstein called this phenomenon "spooky action at a distance" and he was hoping for a quantum physics without such counterintuitive features. On the other hand, there is the novel feature of orbital angular momentum of light, which only came to the attention of the scientific community in the 1990s. The notion of orbital angular momentum carries the possibility that a single photon can carry in principle any amount of angular momentum. In quantum language, in can possess arbitrarily high quantum numbers. The related possibility of the transverse modulation of a light wave, allows novel polarization amplitude and phase patterns to be imprinted on the light beam, even at the single photon level.

It is often said that a transition from quantum to classical phenomena occurs for high quantum numbers. Thus, the potential of a photon to carry arbitrarily large quantum numbers leads, in the eyes of some, to the possibility of studying in detail the quantum to classical transition. In one of his experiments, Robert Fickler demonstrated the entanglement of two photons, each one carrying either $+300\,\hbar$ or $-300\,\hbar$ units of orbital angular momentum. In simple language, this means that neither photon carries any angular momentum until one of them is measured. Then the other one is instantly collapsed into the opposite state. This is certainly the highest number of quanta ever entangled in any experiment. In other words, there is no indication for a transition to classical behavior for these quantum numbers.

To further underline the interesting notion of entanglement, in another experiment Robert Fickler was able to show on a single-photon camera directly in a live movie the change of an orbital angular momentum quantum state of one photon and its dependence on the kind of measurement performed on the other one. This experiment provides a very clear demonstration of the counterintuitive features of quantum entanglement.

Finally, exploiting the possibility of arbitrary transverse modulation of a light beam, Fickler demonstrated the novel, very complex polarization vector and amplitude modulation patterns and their entanglement, again for individual photons. A rather curious result is that two vector photons can be both entangled and not entangled at the same time, depending on the specific measurement performed on them.

Robert Fickler's thesis also opens avenues to novel applications for quantum communication with photons. Particularly the possibility that an individual photon can carry very high quantum numbers allows breaking the barrier that a single photon carries only one bit of information or one qubit of quantum information. Furthermore, his results allow novel quantum cryptography protocols with additional security features.

It is thus evident that Robert Fickler's thesis breaks very new ground in the age-old study of the nature of light and its applications.

June 2015
Vienna

Prof. Anton Zeilinger
University of Vienna and Austrian
Academy of Sciences

Abstract

Quantum entanglement describes the phenomenon where two or more systems show correlations that are stronger than classically explainable, even if they are separated in space. It can be considered as one of the characteristic features of quantum physics and is seen as an important building block for future quantum technologies. In photonic experiments, the transverse spatial degree of freedom offers great potential to explore various interesting features of light. Therefore, to further investigate the nature of quantum entanglement, photonic entanglement of transverse spatial structures has been in the focus of this thesis.

One interesting property due the spatial mode structure of photons is their orbital angular momentum. Surprisingly, there is no theoretical upper limit of how many quanta of orbital angular momentum a single photon can carry. Hence, it appears to be a candidate for testing photonic entanglement of macroscopic values and might contribute to the discussion of macroscopicity and a possible breakdown of quantum mechanics beyond a certain limit. In addition, an increase of the complexity of the spatial-mode structures is often accompanied by unexpected properties, like self-healing or free-acceleration. Thus, an investigation of entanglement of complex structures promises the possibility to discover unseen features of entangled photon pairs.

In this present thesis, we developed methods to generate, investigate, and verify the entanglement of complex spatial structures. Three experiments were performed that highlight the versatility of the introduced schemes for future quantum experiments investigating the spatial degree of freedom of light.

First, we established a novel flexible way to generate spatial-mode entanglement. The scheme enabled the entanglement of two photons with 300 quanta of orbital angular momentum. The results exceed earlier experiments by two orders of magnitude and agree perfectly with quantum mechanical predictions. Our findings represent entanglement of the highest quantum number confirmed in an experiment so far (to my best knowledge) and open a promising route to generate entanglement with even higher quanta. In addition, we have demonstrated that entangled photons with high quanta of angular momentum are beneficial in remote angular sensing applications.

In a second experiment, we introduced a flexible scheme to efficiently observe the spatial mode of the entangled photons with high resolution. With the help of an intensified CCD camera, we imaged the transverse structure of an entangled photon depending on the measurement of its partner photon. The resulting coincidence images have been used to verify the presence of entanglement and to visualize its effect, even in real-time.

In a last experiment, we increased the complexity of the entangled state further by generating a state with transverse spatially varying polarization. Taking advantage of the very high overall efficiency of the whole setup, it was possible to entangle and investigate so-called vector photons (photons where the polarization vector varies depending on the transverse spatial position) in unseen detail. Thereby, we were able to experimentally visualize the different strength of three popular entanglement criteria. Moreover, we demonstrated a novel, surprising property of entangled vector photons: vector photons can be both *entangled* and *not entangled* in polarization depending on their transverse spatial position.

Acknowledgments

First of all, I would like to thank my supervisor Anton Zeilinger. He provided an environment which gave me the opportunity to be part in one of the most interesting research in physics. He made it possible for me to meet many incredibly gifted people, within or outside of physics, and he taught me to ask the important questions and not loosing the bigger picture behind everyday lab work.

I also want to thank all my friends and colleagues, who made that time inspiring, successful, and a lot of fun:

Sven Ramelow, for being a the best graduate-student-post-doc possible and always finding the solutions to unsolvable problems in the lab; Radek Lapkiewcz, for uncovering the weak spots in stupid ideas and having enough great ideas that he cannot realize all of them on his own; Christoph Schäff, for being the best mate and "Erklär-Bär"-fellow that I could imagine through countless dark lab- and pub-hours as well as during explanations in art exhibitions; Mario Krenn, for sharing his inspiring enthusiasm in physics, wuzzeln, Styria, and more; William Plick, for filling my gaps in theoretical physics and showing me the place where it is always sunny; Bernhard Wittmann, for making the time in the artist community unforgettable and for completing the "Erklär-Bär-Trio"; Mehul Malik, for teaching me how to cook $|\text{mild}\rangle + |\text{hot}\rangle$ Indian food; Robert Polster, for bringing Berlin vibes to Vienna; Igor Pikowski, the brilliant mind on which I count for future breakthrough experiments; Armin Hochrainer, the excellent member of the jury in blind testing; Gabi Barreto Lemos, who brought Brazilian atmosphere in our office; Rupert Usrin, Thomas Scheidl, Thomas Herbst, Johannes Handsteiner, Matthias Fink, for sharing all their experience (good and bad) in experimenting outside of the dark lab; Mateusz Kotyrba, party organizer whenever I am away; Max Ebner and Michael Keller, Party and BEC heros; Max Tillman, the guitar hero; Denis Abramovic, Marissa Giustina, Amir Tavala, Johannes Kofler, Sarah Stöckl, Leo Wallin, Amir Moqanaki, Magda Zych, Fabio Costa, Borivoje Dakic, Mateus Santos, and all other colleagues from the quantum groups in Vienna; all people doing the sometimes chaotic administration and making impossible things possible.

I would also like to thank Andrew White for the opportunity to join his awesome group for my CoQuS-secondment in Brisbane; Alessandro Fedrizzi, for his huge hospitality and help with everything; the whole UQ quantum optics group for a fruitful and fun time in Down Under.

I thank Miles Padgett and all the people of his group in Glasgow for a stay that I would not want to miss and that was too short, especially Martin Lavery, for the fun time we spent together whenever and wherever we met.

I also want to thank the University of Vienna, the Austrian Academy of Science and all other fundings for their financial support. None of the results would have been possible without the support through the European Research Council (ERC, QIT4QAT), the European Community Framework Programme 7 (SIQS), the Austrian Science Fund (FWF) within the Special Research Programs (SFB, FoQuS), and the Vienna Doctoral Program on Complex Quantum Systems (CoQuS).

Finally, it was only possible to reach this point with the continuous support of my friends and family. First of all, thank you Regina, my love, for supporting my life as a scientist with all ups and downs conceivable, for being patient whenever the work in the lab rescheduled our plans and most importantly for being you.

Daniel, Robert, and Hüseyin (and families), my best friends that are there now matter what, when and how; the Ulm group (physicists, physicians, biologists…) for making the studies a memorable time and for ongoing friendships. Thanks to all friends that I was lucky enough to spend some time with.

Thank you Claudia, my sister and best friend, for so many reasons too numerous to be named here; Daniel, Jonas, and Leni, for hugs and smiles that only family can give. The whole family in the Allgäu, I would love to see you more often. And most importantly, to my mum Erika and my dad Otto. Thank you for helping me whenever my chaotic time management was out of control, for supporting me in all decisions I made and giving me confidence to belief in the things I am doing—in short: Thank you for being always there for me.

Contents

Chapter 1
Preamble

1.1 Motivation

The physics of light is one of the oldest fields in science and has fascinated many well-known scientists over the last centuries. In the 17th century, a famous contradictory discussion arose when Newton introduced the concept of particles of light, while Huygens described light as a wave-like phenomenon. The latter description was successfully tested in many classic experiments, e.g. the double-slit experiment from Young, and reached its peak popularity in the 19th century. Maxwell formulated his famous description of light as an electromagnetic wave, which might be considered as one of the greatest achievements in physics. However, in the early 20th century, the model was questioned again when Einstein reintroduced the concept of particles of light, nowadays called photons, to explain the photoelectric effect. His explanation can be considered (along with Planck's quantum hypothesis in the description of black body radiation) as one of the starting points of quantum mechanics, the most successful theory we know today. This very brief (oversimplified and incomplete) historical note shows that light has often been in the heart of many fruitful discussions that have yielded a better understanding of physics.

Remarkably, although light has been investigated for a long time, novel properties (even of classical light) are still being discovered today. One especially fruitful property is the angular momentum associated to the spatial mode of the light. In 1992, Allen et al. [1] showed that orbital angular momentum can be attributed to light fields with a spiral phase front, so-called twisted light, and that such phase distribution can be found in a specific family of transverse spatial modes of light, namely Laguerre-Gauss modes. Their seminal paper started a novel, highly active branch in optics, which focuses on the transverse degree of freedom of light. Many optical fields have profited from findings related to this property: optical tweezers use the orbital angular momentum to start a rotating motion [2], optical microscopes take advantage of the spiral phase front to enhance the phase contrast [3], and classical information technologies use the transverse spatial degree of freedom to increase the transmitted data rate [4–6]. Moreover, different transverse light modes have been

R. Fickler, *Quantum Entanglement of Complex Structures of Photons*,
Springer Theses, DOI 10.1007/978-3-319-22231-8_1

investigated and novel interesting properties, like free acceleration of Airy beams [7] or self-healing of Bessel beam structures [8], have been found.

Not only the field of classical optics has profited from a deeper study of the transverse degree of freedom of light. Quantum optical experiments have benefited from the novel properties of a more complex transverse mode structure as well. In their seminal paper in 2001, Mair et al. [9] showed for the first time the entanglement of photons in their orbital angular momentum. Following their idea, many experiments revealed novel features of quantum mechanics: angle-angular momentum uncertainty relations [10, 11], entangled vortex links [12], higher-dimensional entanglement [13–15] and its advantages in applications like cryptography [16] and quantum information [17], to name a few.[1] As in the field of classical optics, different spatial modes and their mode-specific phenomenon, like Bessel-Gauss [19, 20] and Ince-Gauss modes [21], have been explored in the quantum regime.

From both the classical and quantum mechanical experiments described above, an interesting observation can be made: Increasing the complexity of the light field often accompanies the discovery of novel properties of light. Therefore, it might be worthwhile to explore novel approaches to create complex light fields. Moreover, such modes and their properties might be used to test for possible limits of the quantum mechanical description and its features like entanglement. The aim of this thesis is to take a small step in the direction towards answering the following questions:

- Is there a flexible and efficient way to create entanglement between two photons with very complex spatial mode structures?
- Is there a efficient way to observe photonic entanglement of complex mode structures, which is not dependent on a specific mode?

1.2 Outline

The thesis is structured in the following way.

Chapter 2 introduces the theory relevant to the experiments performed and described in this thesis. Starting from classical light fields and their vectorial and spatial properties, a brief explanation of photons as quantum information carriers is given and possible tests to verify entanglement are described. The main purpose of this part is to introduce the required formulae and establish the notation used for later investigations.

Chapter 3 discusses a setup capable of transferring polarization entanglement to entanglement in any transverse spatial mode. The setup was used in a first experiment to generate entangled photon pairs in high-order Laguerre-Gauss modes, which carry very high quanta of orbital angular momentum. Additionally, a simple, novel method to measure such complex states is introduced and used to demonstrate entanglement of the highest quantum numbers so far (to my knowledge).

[1] The interested reader may consult the textbook by Andrews and Babiker [18] for a broader discussion.

A more general and flexible method of detecting spatial mode entanglement is presented in Chap. 4. It is shown that modern imaging technology, namely an intensified CCD camera, is sufficiently fast and efficient to be used in quantum optical experiments. In this experiment, one photon of a polarization-entangled pair was projected to a specific polarization and its detection signal was used as trigger for the intensified CCD camera. The second photon was transferred to a certain spatial mode depending on its polarization and was imaged by the triggered camera. Because of entanglement, any change of the measurement setting of the first trigger photon leads to a different mode structure of the imaged partner photon. Due to the high overall efficiency of all systems, it was possible to visualize the effect of entanglement in real-time. Additionally, the flexibility of the transfer setup and the detection scheme was demonstrated for various mode families.

In a third experiment, described in Chap. 5, the novel creation and detection scheme was extended further. The complexity of the entangled photon state was increased by superimposing different transverse spatial structure each with a different polarization. The resulting light field is commonly called a vector beam, where the polarization varies across the transverse spatial plane of the field. Photons in such complex modes were created and investigated by adapting the earlier transfer setup accordingly and extending the imaging technique to be polarization-sensitive. As in the previous imaging experiment, hybrid-entanglement will be explored. However, this time one photon's polarization was entangled with the other photon's complex polarization pattern. Most importantly, a novel feature of polarization entanglement in vector beams was discovered: photons encoded in specific types of vector beams can be both *entangled* and *not entangled* in polarization depended on the transverse spatial position.

A few summarizing conclusions and a brief outlook for possible future experiments will be given in Chap. 6.

References

1. Allen, L., Beijersbergen, M. W., Spreeuw, R. J. C., & Woerdman, J. P. (1992). Orbital angular momentum of light and the transformation of Laguerre-Gaussian laser modes. *Physical Review A*, *45*, 8185.
2. He, H., Friese, M., Heckenberg, N., & Rubinsztein-Dunlop, H. (1995). Direct observation of transfer of angular momentum to absorptive particles from a laser beam with a phase singularity. *Physical Review Letters*, *75*, 826.
3. Fürhapter, S., Jesacher, A., Bernet, S., & Ritsch-Marte, M. (2005). Spiral phase contrast imaging in microscopy. *Optics Express*, *13*, 689.
4. Tamburini, F., et al. (2012). Encoding many channels on the same frequency through radio vorticity: first experimental test. *New Journal of Physics*, *14*, 033001.
5. Wang, J., et al. (2012). Terabit free-space data transmission employing orbital angular momentum multiplexing. *Nature Photonics*, *6*, 488.
6. Bozinovic, N., et al. (2013). Terabit-scale orbital angular momentum mode division multiplexing in fibers. *Science*, *340*, 1545.
7. Siviloglou, G. A., Broky, J., Dogariu, A., & Christodoulides, D. N. (2007). Observation of accelerating airy beams. *Physical Review Letters*, *99*, 213901.

8. Broky, J., Siviloglou, G. A., Dogariu, A., & Christodoulides, D. N. (2008). Self-healing properties of optical airy beams. *Optics Express*, *16*, 12880.
9. Mair, A., Vaziri, A., Weihs, G., & Zeilinger, A. (2001). Entanglement of the orbital angular momentum states of photons. *Nature*, *412*, 313.
10. Leach, J., et al. (2010). Quantum correlations in optical angle-orbital angular momentum variables. *Science*, *329*, 662.
11. Jack, B., et al. (2011). Demonstration of the angular uncertainty principle for single photons. *Journal of Optics*, *13*, 064017.
12. Romero, J., et al. (2011). Entangled optical vortex links. *Physical Review Letters*, *106*, 100407.
13. Vaziri, A., Weihs, G., & Zeilinger, A. (2002). Experimental two-photon, three-dimensional entanglement for quantum communication. *Physical Review Letters*, *89*, 240401.
14. Dada, A. C., Leach, J., Buller, G. S., Padgett, M. J., & Andersson, E. (2011). Experimental high-dimensional two-photon entanglement and violations of generalized bell inequalities. *Nature Physics*, *7*, 677.
15. Krenn, M., et al. (2014). Generation and confirmation of a (100 × 100)-dimensional entangled quantum system. *Proceedings of the National Academy of Sciences*, *11*, 6243.
16. Gröblacher, S., Jennewein, T., Vaziri, A., Weihs, G., & Zeilinger, A. (2006). Experimental quantum cryptography with qutrits. *New Journal of Physics*, *8*, 75.
17. Langford, N. K., et al. (2004). Measuring entangled qutrits and their use for quantum bit commitment. *Physical Review Letters*, *93*, 053601.
18. Andrews, D . L., & Babiker, M. (2012). *The angular momentum of light*. Cambridge: Cambridge University Press.
19. McLaren, M., et al. (2012). Entangled Bessel-Gaussian beams. *Optics Express*, *20*, 23589.
20. McLaren, M., Mhlanga, T., Padgett, M. J., Roux, F. S., & Forbes, A. (2014). Self-healing of quantum entanglement after an obstruction. *Nature Communications*, *5*, 3248.
21. Krenn, M., et al. (2013). Entangled singularity patterns of photons in Ince-Gauss modes. *Physical Review A*, *87*, 012326.

Chapter 2
Introduction and Theoretical Background

2.1 Preamble

The nature of light and its main properties, either seen classically as an electromagnetic wave or quantum mechanically as a photon, is not only the focus of much ongoing research, but has been covered by many excellent textbooks. Thus, only a short introduction to the theory will be conducted in the following chapter mainly to establish the employed nomenclature.

2.2 Light as Electromagnetic Waves

I will follow essentially the discussions and calculations of the standard text books in optics of Born and Wolf, Saleh and Teich and Siegman [1–3], the textbook of Andrews and Babiker [4] and the PhD-thesis of A. Mair [5], adapting them to the nomenclature used in this thesis.

2.2.1 Paraxial Wave Equation

The most basic and general way to describe classically the phenomena of light is the theory of electromagnetic fields. Here, the basic equations are the Maxwell equations, which fully describe all phenomena known to classical light fields as electromagnetic waves. However, in this thesis, a simpler but still powerful approximation can be used as the basic theoretical background, the so-called paraxial approximation. The time-independent complex amplitude of the electromagnetic wave can be written as

$$U(\mathbf{r}) = u(\mathbf{r}) \exp(ikz), \tag{2.1}$$

R. Fickler, *Quantum Entanglement of Complex Structures of Photons*,
Springer Theses, DOI 10.1007/978-3-319-22231-8_2

where the vectorial character of light is assumed to be uniform (and will be disregarded for the moment), z is the direction of propagation and $\mathbf{r}$ describes the three-dimensional coordinate in space. The simplifying assumption is now that the field varies slowly with respect to the propagation direction z (within a wavelength λ: $\Delta u \ll u$). Hence, the complex amplitude can be considered as a scalar wave with a "modulated" envelope u (herein called spatial mode), which travels in z-direction. From this approximation the *paraxial wave equation* (PWE) can be derived[1]

$$\left(\Delta_\perp + 2ik\frac{\partial}{\partial z}\right) u(\mathbf{s}, z) = 0, \tag{2.2}$$

where $\Delta_\perp$ represents the transverse Laplacian operator and $\mathbf{s}$ labels the position in the transverse plane in the corresponding coordinate system (e.g. (x, y) for Cartesian, (r, θ) for cylindrical and (u, v) for elliptical coordinates).

It can be shown that the paraxial approximation is justified if the focusing and diverging angles of the light field are smaller than approximately 30° before notable corrections have to be taken into account [3]. Although the PWE (2.2) accurately describes all experimental investigations presented in this thesis, one should always bear in mind that it is only an approximation. Moreover, it was pointed out in detail by Lax et al. [6] that the paraxial approximation, applied to the Helmholtz equation for electric vectorial fields leads to a contradiction. For transversely-polarized fields, i.e. the electric field oscillates transverse to the optical axis, the transverse directional derivative of the electric field is equal to zero. Thus, the transverse Laplacian $\Delta_\perp$ is also equal to zero, which contradicts paraxial assumption. It was shown by Lax et al. [6] (and later by Davis [7] in a simpler way for vector potentials) that the paraxial approximation can be naturally obtained from lowest-order terms by scaling the Maxwell equations and expanding the field appropriately.

Fields outside of this simplified description might be a future research direction to investigate unknown properties of light fields. Although outside of the scope of this thesis, the interesting approach from Bialynicki-Birula introducing the Riemann-Silberstein vector [8] might be an avenue to investigate the full properties of light described by the Maxwell equations. In the following sections, a detailed discussion of the transverse spatial modes, the polarization of the field and their connection to the angular momentum of light will be conducted. After that, it will be shown that non-trivial combinations of both lead to interesting, complex modes of light.

2.2.2 Transverse Spatial Modes

The PWE (2.2), which was derived in the last section, can be solved in different coordinate systems, each resulting in a different spatial mode family. Since the solution

[1] The derivation from the Maxwell equations can be found for example in the textbook [4] or the thesis of A. Mair [5].

is restricted to the paraxial regime, where the z-axis is the Cartesian axis of propagation, only solutions for different, orthogonal coordinate systems in the 2-dimensional transverse plane are considered. The coordinate systems can be classified depending on their polynomial degree [9] and the separability of their coordinates. This leads to different spatial modes (a more detailed characterization can be found in the Masters's thesis of M. Krenn [10]). In the following, only coordinate systems with polynomial degree one or two will be taken into account, which can be fully separated, i.e. solved by the separation of the variable ansatz. Furthermore, only the elliptical coordinate system rather than the parabolic coordinates will be investigated (although solutions have been found [11] and experimentally demonstrated [12]). In the next section two special cases of the solutions will be described first: Cartesian coordinates [13], from which the known Hermite-Gauss beams can be derived, and circular coordinates [14], which result in the mode family of Laguerre-Gauss beams. After that, the general solutions of elliptical coordinates systems [15] and their mode family of Ince-Gauss beams [16] will be discussed.

Hermite-Gauss Modes

If the PWE (2.2) is solved in Cartesian coordinates $u(x, y, z)$, the complete orthogonal set of Hermite-Gauss (HG) modes is obtained[2]:

$$HG_{m,n} := u_{m,n}(x, y, z) = \frac{\sqrt{2}}{\sqrt{\pi m! n! 2^{(m+n)}}} \frac{1}{w(z)} H_m\left(\sqrt{2}\frac{x}{w(z)}\right) H_n\left(\sqrt{2}\frac{y}{w(z)}\right)$$
$$\exp\left[\frac{ik(x^2+y^2)}{2R(z)} - \frac{(x^2+y^2)}{w^2(z)}\right] e^{i\phi_{m,n}}, \quad (2.3)$$

where $w(z) = w_0(1 + (z/z_r)^2)^{1/2}$ is the beam diameter at the z-position, $R(z) = z(1+z_r^2/z^2)$ is the radius of curvature of the mode, $z_r = \frac{\pi w_0^2}{\lambda}$ is the Rayleigh length, and w_0 is the beam waist.

The Hermite polynomials, which give the mode family its name, are defined as

$$H_m(x) = (1)^m \exp(x^2) \frac{d^m}{dx^m} \exp(-x^2), \quad (2.4)$$

where $m \in \mathbb{N}$ (analogous for H_n) and the Gouy phase of Hermite-Gauss modes is

$$\phi_{m,n} = \left(1 + \frac{m+n}{2}\right) \arctan(z/z_r). \quad (2.5)$$

The Gouy phase describes the change of the phase when passing through the focus (point of minimal beam waist w_0) and might be used to classify different mode "generations" [18]. Superpositions built by modes of the same generation, i.e. modes with the same Gouy phase, are stable under z-propagation (the transverse spatial structure does not change besides the natural spreading of the beam). In Fig. 2.1, the

[2] A detailed derivation can be found e.g. in [3, 17].

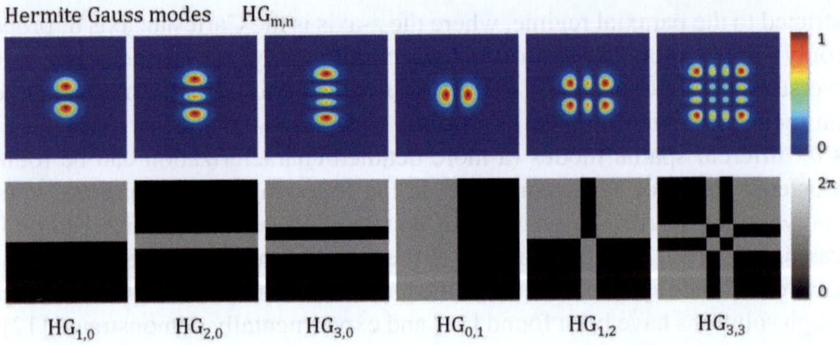

Fig. 2.1 Hermite-Gauss mode gallery. The *upper row* shows calculated intensity structures of various HG modes (normalized). In the *lower row*, the corresponding phase structures can be found. The π-phase steps correspond to the two different characteristic numbers m and n. The first number m is related to the amount of *horizontal* phase steps, the second n to *vertical* ones. This phase discontinuities can be seen as nodal lines in the intensity structure

intensity structure and phase distribution are shown for different HG modes. It can be seen that the characteristic numbers m and n correspond to horizontal and vertical π-phase steps, which leads to nodal lines in the intensity profile.

Laguerre-Gauss Modes

Another family of transverse spatial modes can be found if the PWE (2.2) is solved in cylindrical coordinates $u(r, \theta, z)$, where r is the radius and θ stands for the azimuthal angle. Those rotationally symmetric modes are called Laguerre-Gauss modes and represent another complete, orthogonal set of spatial modes.

$$LG_{l,n} := u_{l,n}(r,\theta,z) = \frac{\sqrt{2n!}}{\sqrt{\pi(|l|+n)!}} \frac{1}{w(z)} \left(\frac{\sqrt{2}r}{w(z)}\right)^{|l|} L_n^{|l|}\left(\frac{2r^2}{w^2(z)}\right) \cdot \exp\left[\frac{ikr^2}{2R(z)} - \frac{r^2}{w^2(z)}\right] e^{i\phi_{l,n}} e^{il\theta}, \tag{2.6}$$

where the beam diameter $w(z)$, the radius of curvature $R(z)$, and the Rayleigh length are defined analogously as for HG modes. The Laguerre polynomials are

$$L_n^l(r) = \frac{e^r r^{-l}}{n!} \frac{d^n}{dr^n}\left(e^{-r} r^{l+n}\right), \tag{2.7}$$

with $l \in \mathbb{Z}$ and $n \in \mathbb{N}$. Similarly to HG modes, LG modes a have mode dependent Gouy phase $\phi_{l,n}$, which is

$$\phi_{l,n} = (|l| + 2n + 1) \arctan(z/z_r). \tag{2.8}$$

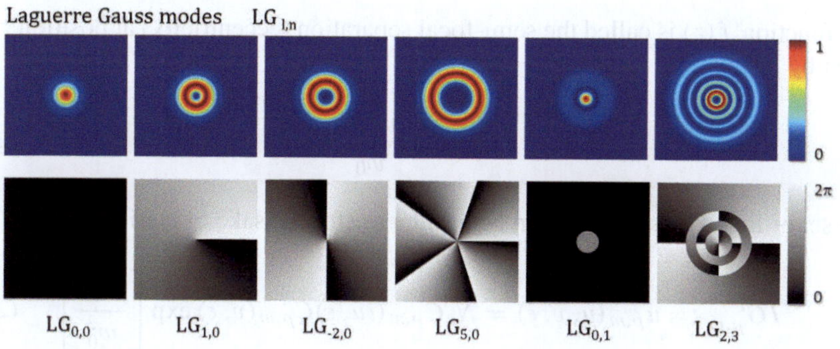

Fig. 2.2 Laguerre-Gauss mode gallery. The *upper row* shows calculated intensity structures (normalized), the *lower row* the corresponding helical phase distribution. The azimuthal phase dependence corresponds to the first characteristic number l and is directly related to the orbital angular momentum of the light (see Sect. 2.2.4). The phase singularity in the center leads to a vanishing intensity along the optical axis. The second number n is related to the number of radial phase steps, which can be seen as nodal rings in the intensity structure

Again, the Gouy phase can be used the formulate different mode generations, where superpositions of the generation have a stable transverse beam profile when propagating. LG mode superpositions that are built by modes with different Gouy phase rotate during propagation.[3] The intensity structure and phase distribution of various LG modes can be seen in Fig. 2.2. The two characteristic numbers l and n describe the azimuthal phase structure and the number of radial π-phase steps. If l is unequal to zero, the phase forms a helical structure. The steepness of helical phase front is defined by the number l, which is related to the orbital angular momentum of light and is called the topological charge (see Sect. 2.2.4). As a result of the screw like phase structure, these beams exhibit a phase singularity in the center. Thus, they have an intensity null along the optical axis. For higher orders of n, nodal rings can be found in the intensity structure.

Due to the angular momentum associated with LG modes, they have attracted a lot of attention over the last 20 years. For a more detailed discussion of these modes, i.e. their properties as well as recent applications, the interested reader is directed to the recent book by Andrews and Babiker [4] or the review article by Franke et al. [19].

Ince-Gauss Modes

A more general mode family is constituted by the Ince-Gauss (IG) modes [16]. They are solutions to the PWE in elliptical coordinates $u(u, v, z)$, which are related to Cartesian coordinates as follows

$$\begin{pmatrix} x \\ y \end{pmatrix} = f(z) \begin{pmatrix} \cosh(u)\cos(v) \\ \sinh(u)\sin(v) \end{pmatrix}. \tag{2.9}$$

[3] A more detailed description can be found here [18].

The function $f(z)$ is called the semi-focal separation (eccentricity) at position z and given by

$$f(z) = f_0 \frac{w(z)}{w_0}. \tag{2.10}$$

The set of Ince-Gauss modes can be written as (for the sake of clarity $z = 0$)

$$IG^e_{p,m} := u_{p,m}(u, v, \epsilon) = N_e C_{p,m}(iu, \epsilon) C_{p,m}(v, \epsilon) \exp\left[\frac{-r^2}{w_0^2}\right] \tag{2.11}$$

$$IG^o_{p,m} := u_{p,m}(u, v, \epsilon) = N_o S_{p,m}(iu, \epsilon) S_{p,m}(v, \epsilon) \exp\left[\frac{-r^2}{w_0^2}\right], \tag{2.12}$$

where $C_{p,m}$, $S_{p,m}$ are the Ince polynomials, $p, m \in \mathbb{N}$, the ellipticity parameter is $\epsilon = 2f_0^2/w_0^2$, N_o, N_e are normalization constants, and e, o label the even and odd Ince-Gauss modes. The radial-cylindrical coordinate r is defined as

$$r = f_0 \sqrt{\cosh^2(u)\cos^2(v) + \sinh^2(u)\sin^2(v)}. \tag{2.13}$$

A detailed discussion about solving the PWE in elliptical coordinates can be found in the seminal paper of Bandres and Vega [16] or the Masters thesis of Mario Krenn [10]. The superposition of the even and odd Ince-Gauss modes (Eqs. (2.11) and (2.12)) leads to the helical Ince-Gauss modes (herein called "Ince-Gauss" modes)

$$IG^{\pm}_{p,m} := u_{p,m}(u, v, \epsilon) = \frac{1}{\sqrt{2}} \left(u^{IGe}_{p,m}(u, v, \epsilon) \pm i u^{IGo}_{p,m}(u, v, \epsilon) \right). \tag{2.14}$$

The two characteristic numbers of Ince-Gauss modes p and m are restricted to have the same parity, i.e. both are even or both are odd, and to fulfill $m \leq p$. Various types of these modes can be found in Figs. 2.3 and 2.4.

In Fig. 2.3, where modes for a fixed ellipticity are shown, it can be seen that m corresponds to the number of hyperbolic nodal lines in the intensity structures. The number of elliptic nodal lines is given by the number $(p - m)/2 + \gamma$, where γ is 1 for odd modes and 0 for even modes. In Fig. 2.4, modes with fixed m and p are presented while the ellipticity is continuously increased. If the ellipticity ϵ is zero, the IG modes become LG modes with the topological charge $l = m$ and $n = (p - m)/2$ radial nodes. In the limit of infinite ellipticity $\epsilon \to \infty$, the IG modes become helical Hermite-Gauss modes [20] (which can be decomposed into two HG modes). Another interesting feature is the continuous splitting of the vortices with increasing ellipticity. Depending on the topological charge of the vortex if $\epsilon = 0$ (the quantum number m), the modes show splitting into m vortices along the transverse horizontal axis if the ellipticity is increased.

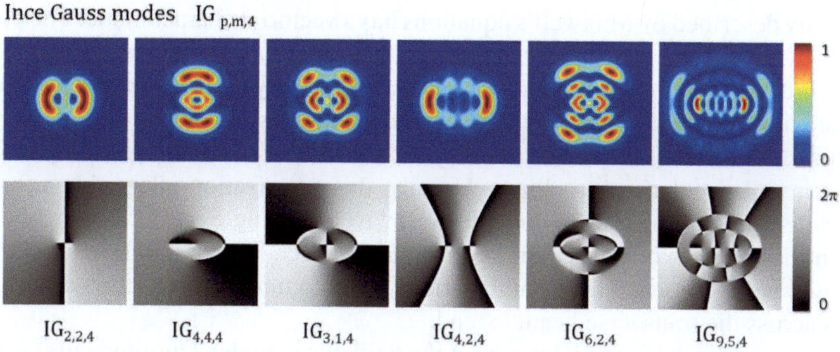

Fig. 2.3 Ince-Gauss mode gallery for ellipticity $\epsilon = 4$. Analogous to the previous figures, the calculated intensity structures and the corresponding phase structure is shown in the *upper* and *lower row*. The second number m is related to *hyperbolic nodal lines* intensity structure, whereas from a combination of p and m the *elliptical nodal lines* can be deduced (formula given in the main text)

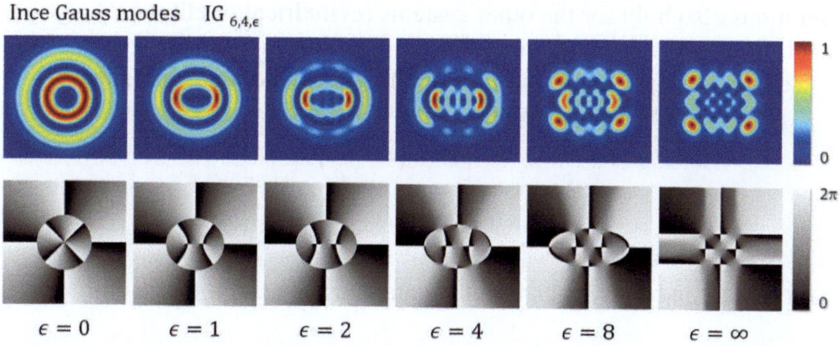

Fig. 2.4 Ince-Gauss mode gallery for one mode with increasing ellipticity. The *upper* and *lower row* demonstrate the change in the intensity and phase structure. If the ellipticity equals zero, the IG mode becomes a LG mode. If the ellipticity goes to infinity, the IG mode turns into a helical HG mode

Similarly to LG modes, an orbital angular momentum can also be assigned to IG modes. A detailed discussion of the OAM value per photon can be found in [10, 21] and reveals a continuously varying OAM (incl. increasing and decreasing fractional OAM values) depending on the ellipticity.

Before the angular momentum connected to electromagnetic fields is discussed in Sect. 2.2.4, a brief investigation of another important property of light is conducted in the following section, namely the vectorial nature of light fields i.e. the polarization.

2.2.3 Polarization

So far, the classical light field was considered to be only a transverse, scalar field $u(\mathbf{s}, z)$. However, as stated earlier this is just an approximation. The electromagnetic field fully described by Maxwell's equations has a vectorial characteristic. The polarization of the light is such a vectorial property. It is determined by the temporal change of the direction of the electromagnetic field and might vary for different positions **r**. In the paraxial approximation the field vector lies in the transverse plane (spanned by the vectors **s**). The curve that is described by the movement of the endpoint of the field vector is called polarization ellipse. Usually, this polarization ellipse is uniformly distributed in the paraxial regime. However, in following Sect. 2.2.5 the coherent superpositions of two different transverse modes, each having a different polarization, will be discussed. These combinations leads to modes where the polarization varies across the transverse beam extend.

To account for the vectorial nature of the field, the complex amplitude $\mathbf{u}(\mathbf{s})$ is now described as a vector field and written in the Cartesian coordinate system as

$$\mathbf{u} = u_x \mathbf{x} + u_y \mathbf{y}. \tag{2.15}$$

Note that it is most convenient to use the Cartesian coordinate system although all considerations also hold for the other systems (cylindrical or elliptical).

The electric field $\mathbf{E}(z, t)$ that travels in z-direction can now be separated into its transverse components (for the sake of simplicity a plane wave is assumed)

$$E_x = u_x \cos\left[2\pi\nu\left(t - \frac{z}{c}\right) + \delta_x\right] \tag{2.16a}$$

$$E_y = u_y \cos\left[2\pi\nu\left(t - \frac{z}{c}\right) + \delta_y\right]. \tag{2.16b}$$

These formulas can be rewritten into the parametric equation of the ellipse

$$\frac{E_x^2}{u_x^2} + \frac{E_y^2}{u_y^2} - 2\frac{E_x}{u_x}\frac{E_y}{u_y}\cos\delta = \sin^2\delta, \tag{2.17}$$

where the phase difference between the two components is $\delta = \delta_x - \delta_y$.

Different Types of Polarization

The state of the polarization is described by the ellipse of Eq. (2.17). Hence, *elliptically-polarized* light is defined by three parameters: the amplitudes u_x and u_y and the phase difference δ (see Fig. 2.5). However, there are two special cases, which are of particular interest (especially because they are often used as a basis to describe the polarization state in the quantum regime, see Sect. 2.3.2).

The first important polarization state arises when the phase difference is $\delta = \pm\pi/2$ and both amplitudes are equal $u_x = u_y$. As seen in Eq. (2.17), the ellipse turns into a circle. Thus, the light is *circularly polarized*. The common convention is that if $\delta = +\pi/2$ the light is said to be right handed circularly polarized (at a fixed position, the electric field rotates clockwise around the optical axis z if seen from the plane that the light approaches). For $\delta = -\pi/2$, the light is called left handed circularly polarized.

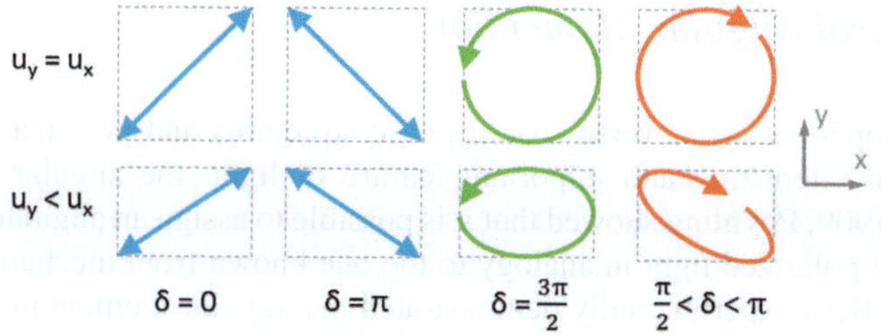

Fig. 2.5 Examples of different polarizations depending on the amplitudes u_x and u_y and phase difference δ. Throughout the whole thesis, the color coding presented here, will be used. Linear polarization is depicted in *blue* (first two columns), left handed circular or elliptical in *green* (third column), and right handed circular or elliptical in *red* (fourth column)

The second special case is the *linear polarization*. If the phase difference δ is either 0 or π, the ellipse collapses to a straight line. The slope of the linear polarization depends on the ratio of $\pm u_y/u_x$, where the sign corresponds to the phase difference δ of 0 or π. Following the standard convention for linearly-polarized light, the polarization will be called horizontal (H), vertical (V), diagonal (D) or anti-diagonal (A) if $u_y = 0$, $u_x = 0$, $\frac{u_y}{u_x} = +1$ or $\frac{u_y}{u_x} = -1$, respectively.

Another convenient description of the polarization state of the light are the so-called Stokes parameters. They can easily be measured (see Sect. 5.1.2) and are related to the amplitudes and phases in the following way:

$$S_0 = u_x^2 + u_y^2 \tag{2.18a}$$

$$S_1 = u_x^2 - u_y^2 \tag{2.18b}$$

$$S_2 = 2u_x u_y \cos\delta \tag{2.18c}$$

$$S_3 = 2u_x u_y \sin\delta. \tag{2.18d}$$

Here, S_0 depicts the intensity of the light and the parameters S_1, S_2, and S_3 correspond to the ratio of the amplitudes, the ellipticity, and orientation of the electromagnetic light field. For a fully polarized beam, the Stokes parameters can be considered as a real three-dimensional vector in Cartesian coordinates, which can be assigned to spherical coordinates

$$\begin{pmatrix} S_1 \\ S_2 \\ S_3 \end{pmatrix} = S_0 \begin{pmatrix} \cos(2\epsilon)\cos(2\theta) \\ \cos(2\epsilon)\sin(2\theta) \\ \sin(2\epsilon) \end{pmatrix}, \tag{2.19}$$

where the ellipticity ϵ ranges from $-\pi/4$ to $\pi/4$ and the orientation θ can take any value between 0 and π. The sphere, which is spanned by all possible configuration (S_0 constant) of ϵ and θ, is called the Poincaré sphere. Every point on its surface corresponds to a specific polarization. Linear polarizations can be found on the equator, circular polarizations on the poles, and elliptical polarizations everywhere between (see Fig. 2.11 for more information).

2.2.4 Optical Angular Momentum

The two properties—transverse spatial phase structures and polarization—can be linked to an interesting and important feature of light: the angular momentum. Already in 1909, Poynting showed that it is possible to assign an angular momentum to circularly-polarized light in analogy to the one known from mechanical motions [22]. Later, Beth experimentally demonstrated the angular momentum of light. He showed that it can be understood not only classically but quantum mechanically by assigning $\pm\hbar$ of angular momentum to the circularly-polarized photons as a spin angular momentum (SAM) [23]. In 1992, Allen, Woerdmann, and coworkers discovered that an orbital angular momentum (OAM) can be identified, if the light field has a helical phase structure, e.g. the phase term $\exp(il\theta)$ in LG modes [24]. Depending on the inclination of the screw-like phase structure, thus depending on the l value, l quanta of OAM can be assigned to each photon. This type of angular momentum was later demonstrated for the first time in experiments by the group of Rubinstein-Dunlop [25, 26], where particles were set in rotation. Intuitively (and seen in experiments), the difference between the two angular momenta can be understood as follows: if circularly-polarized light is absorbed from trapped particles, they start to spin around their own axis, whereas OAM containing light fields, e.g. LG modes, lead to a rotation of the particles around the optical axis [27].

Classically, a straightforward separation between the two angular momenta arises, if the total angular momentum of light is considered

$$\mathbf{J} = \int \mathrm{d}V \epsilon_0 \mu_0 \mathbf{r} \times (\mathbf{E} \times \mathbf{H}). \tag{2.20}$$

This can be transformed with the vector potential $\mu_0\mathbf{H} = \nabla \times \mathbf{A}$ and Gauss' theorem (quickly declining electromagnetic fields assumed) to

$$\mathbf{J} = \int \mathrm{d}V \epsilon_0 E_i (\mathbf{r} \times \nabla) A_i + \int \mathrm{d}V \epsilon_0 \mathbf{E} \times \mathbf{A} \tag{2.21}$$

$$= \mathbf{L} + \mathbf{S}, \tag{2.22}$$

where the analogy to the quantum mechanical angular momentum operator suggests that $\mathbf{L}$ is related to the orbital part of the angular momentum. The dependence of $\mathbf{S}$ on the vectorial nature of the light field, i.e. the polarization, assigns the second term to the spin part of the angular momentum.

Despite this seemingly obvious separation into SAM and OAM, there has been a lot of discussions and confusions about its meaning. The confusion mainly arose due to the fact that neither of the two momenta itself, regarded as quantum mechanical operators $\hat{\mathbf{L}}$ and $\hat{\mathbf{S}}$, fulfills the commutation relation for angular momenta (for more details see [4, 28] and citations therein). However, for paraxial beams it was shown that the separation into SAM and OAM can be done in a meaningful way [29], hence paraxial beams have a well defined spin and orbital angular momentum in the direction of propagation. Moreover, in paraxial beams the defining operators $\hat{S}_z$ and $\hat{L}_z$ fulfill the commutation relation for angular momentum.

This is somewhat remarkable because it implies that single photons are not limited to one quantum of angular momentum. Theoretically, single photons can have arbitrary large angular momentum values, e.g. $l \cdot \hbar$ of OAM for LG modes depending on the phase factor $\exp(il\theta)$. This result is of great importance for experiments presented later, where the entanglement of high angular momenta is investigated.

For reasons of completeness, it is worthwhile mentioning that modes can carry fractional OAM as well, e.g. IG modes [21] or beams with a helical phase structure that is not a multiple of 2π (non-integer values of l) [30, 31].

2.2.5 Vector Beam Families

In the previous sections, either the transverse spatial mode structure was investigated assuming the same polarization across the whole beam or polarization was discussed disregarding the modal structure of the beam. In the following section, the discussions will be generalized by investigating a superposition of two different transverse spatial modes, that are orthogonally polarized to each other. The common feature of such superpositions is a transverse varying polarization, which led to their name "vector beams" [4, 32].

A general vector beam can be described by

$$\mathbf{Vec} = \frac{1}{\sqrt{2}}\big(u_{spM}(\mathbf{s}) \cdot \mathbf{e}_{pol} + e^{i\varphi} u_{spM}(\mathbf{s}) \cdot \mathbf{e}_{pol}\big), \tag{2.23}$$

where u_{spM} labels different spatial modes, $\mathbf{e}_{pol}$ is the unit vector related to a specific polarization, and φ can take any value between 0 and 2π. For the sake of simplicity, the considerations will be restricted to vector modes at the position $z = 0$. Specific features connected to propagation will be mentioned when appropriate (e.g. Poincaré beams).

Similar to the spatial mode families from Sect. 2.2.2, they can be classified into different families depending on their type polarization pattern. The pattern is naturally linked to the spatial modes and polarizations used to construct the vector modes. In the following sections, only vector beams constructed by zeroth or first-order spatial modes are considered, although vector beams can also be built by higher-order modes [33, 34]. However, they do not show other properties than first-order vector beams, apart from a higher frequency of the polarization change across the beam extent (higher-order symmetries).

Cylindrical Vector Beams

An frequently studied family of vector beams is that of so-called "cylindrical vector beams", which have a cylindrically symmetric polarization pattern [32]. The large interest in these beams can be attributed to different applications of which a few are mentioned in the following. It was shown, for example that such beams, more specifically radially-polarized beams, can increase the excitation rate of surface-plasmons [35, 36]. Moreover, they can be used to focus light beams to smaller spots [37, 38], which increases the resolution in microscopy applications. In addition, they allow photon-atom coupling efficiencies up to 100 % [39, 40], which could be very beneficial if quantum networking tasks between photons and single atoms are realized.

Cylindrical vector beams can be constructed by superposing either linearly-polarized HG modes or circularly-polarized LG modes. For an easier distinction, the first types will be called "HG vector beams", while latter ones will be named "LG vector beams".

HG Vector Beams

One way to construct cylindrical vector beams is to superpose two HG modes with linear polarization. Hence, Eq. (2.23) can be rewritten as

$$\mathbf{Vec}^{HG} = \frac{1}{\sqrt{2}}\big(HG_{1,0} \cdot \mathbf{e}_{H/V} + e^{i\varphi} HG_{0,1} \cdot \mathbf{e}_{V/H}\big), \tag{2.24}$$

following the nomenclature introduced in the previous sections. The arising polarization pattern of first-order HG vector beams for different phases φ can be seen in Fig. 2.6 (columns three to six). For specific superpositions, the popular cylindrical vector beams with azimuthal (1st row, $\varphi = 0°$) or radial polarization (2nd row, $\varphi = 0°$) is found. The so-called anti-vortex configuration (1st and 2nd row, $\varphi = 180°$) shows a rotation of the linear polarization vector contrary to its angular position.

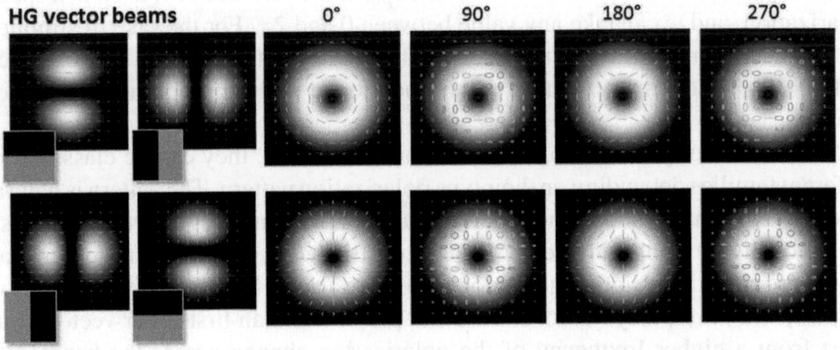

Fig. 2.6 Polarization pattern of HG vector beams. The calculated intensity is shown in *gray scale* in the background and is overlaid by the vector beam polarization pattern. The shapes and colors of the polarization vectors depict the type of polarization (color coding as introduced in Fig. 2.5: *blue* = linear, *red* = right elliptic, *green* = left elliptic). The components out of which the vector beams are constructed are shown in the first two columns (small insets depict the phase of the beam)

LG vector beams

A second way to construct cylindrical vector beams is the superposition of circularly-polarized LG modes

$$\mathbf{Vec}^{LG} = \frac{1}{\sqrt{2}}\Big(LG_{1,0}\cdot \mathbf{e}_{R/L} + e^{i\varphi} LG_{-1,0}\cdot \mathbf{e}_{L/R}\Big). \tag{2.25}$$

The polarization pattern of LG vector beams composed of circularly-polarized $LG_{1,0}$ and $LG_{-1,0}$ is visualized in Fig. 2.7. All polarization pattern in the upper row have only linear polarization vectors with the cylindrical symmetric azimuthal ($\varphi = 180°$), radial ($\varphi = 0°$), and spiral ($\varphi = 90°, 270°$) configuration. The combination depicted in the lower row leads to the anti-vortex structure, where different phases φ result in a rotation of the polarization pattern.

With both types, LG and HG vector beams, radially and azimuthally-polarized vector beams can be generated for different configurations (described above). However, the difference between the two types is clearly visible (see Figs. 2.6 and 2.7). The pattern of LG vector beams exhibit only linear polarizations in all superpositions. Another crucial difference becomes apparent, if one compares specific transverse positions throughout one type of superposition (one row in Fig. 2.6 or 2.7). For HG vector beams, the horizontal and vertical central regions always show the same polarization, whereas for LG vector beams any transverse position has a different polarization if the phase changes. This feature can be explained intuitively by the modal structure of the spatial components (HG or LG modes). At regions where no modal overlap occurs (e.g. for HG vector beams at horizontal and vertical central regions), the polarization is independent of the phase φ. This special characteristic is important and will be discussed later in more detail, when entanglement of such modes is investigated.

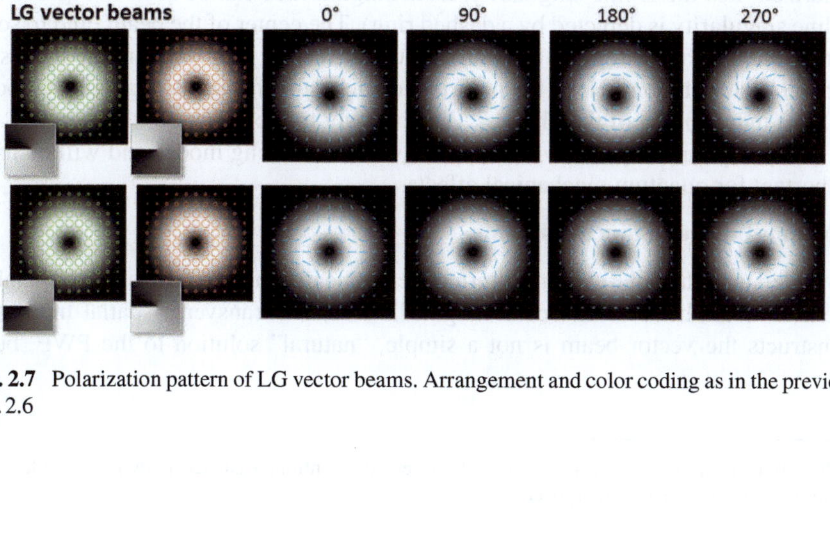

Fig. 2.7 Polarization pattern of LG vector beams. Arrangement and color coding as in the previous Fig. 2.6

Poincaré Beams

Only recently, a novel class of vector beams containing every polarization on the Poincaré sphere was presented [41] (a short discussion including their specific polarization patterns can already be found in the PhD thesis of Mair [5]). These so-called "Poincaré beams" have attracted much attention, because they exhibit interesting features like different types of polarization singularities and a change of the polarization pattern while freely propagating. Polarization singularities are points or lines in the transverse polarization pattern, where a property of the polarization vector is undefined. In Poincaré beams, so-called C-point singularities, i.e. points where the orientation of the polarization ellipse is undefined (circular polarization) and L-line singularities, where the handedness of the polarization is undefined (linear polarization), can be found [34, 42–46]. Another interesting effect, the rotation of the pattern during free propagation, can be explained by the different Gouy phases of the two superposed modes[4] [47].

The simplest way to construct Poincaré beams is to superpose a circularly-polarized Gauss mode with an orthogonally-polarized LG mode of the first order

$$\mathbf{Vec}^P = \frac{1}{\sqrt{2}} \left(LG_{0,0} \cdot \mathbf{e}_{R/L} + e^{i\varphi} LG_{1,0} \cdot \mathbf{e}_{L/R} \right). \tag{2.26}$$

The polarization pattern of Poincaré beams (Fig. 2.8) exhibits every polarization, starting with circular polarization in the center and continuously changing to the orthogonal circular polarization depending on the radial position. Depending on the azimuthal position, the slope of the elliptical (or linear) polarization varies through all possible angles. Hence, the whole Poincaré sphere is stereographically projected onto the real, physical plane of the transverse light field. Furthermore, the pattern of Poincaré beams contains different polarization singularities. In the center of the beam a C-point singularity is located. Around the center, where the light is linearly polarized, lies the L-line singularity. Both singularities can be seen in Fig. 2.8 (the L-line singularity is depicted by a dashed ring). The center of the beam (and to some extent the outermost regions) has an additional feature that was already discussed for HG vector photons. Here, the polarization does not change throughout all modes, i.e. it is independent of the phase φ. Analogously to the HG vector beams, this can be easily explained by the overlap of the two constructing modes and will be more important for quantum mechanical effects.

Custom-Tailored Poincaré Beams

To understand the polarization pattern of vector beams more intuitively, "artificially" created vector beams will be investigated. Here, the transverse spatial mode that constructs the vector beam is not a simple, "natural" solution to the PWE, but a

[4]This effect is similar to the rotation of the intensity mentioned earlier, if two LG modes with different Gouy phase are superposed.

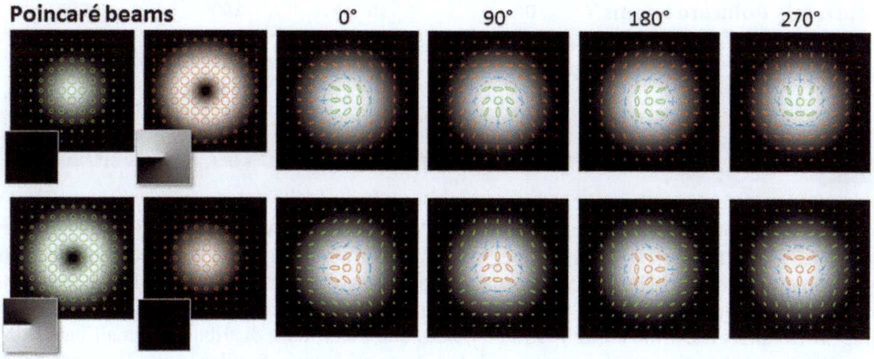

Fig. 2.8 Polarization patterns of Poincaré beams. Again, the background shows the calculated intensity in *gray scale* and is overlaid by the polarization pattern (shapes and colors of the polarization vectors depict the type of polarization; color coding as in Fig. 2.6). On the *left side* (first two columns), the components to build the Poincaré beams are shown including the corresponding phase distribution (small insets). The *upper row* shows the so-called lemon configuration, the *lower row* demonstrates a star configuration (following the characterization of surface singularities—umbilic points—by Berry and Hannay [48]). Both patterns exhibit a C-point polarization singularity in the center and L-line singularities around the optical axis (*dashed circle*)

square shaped beam.[5] Within the square shaped mode, the intensity varies linearly from left to right. For one mode it increases from zero to maximum intensity, for the other mode it decreases from maximum to zero intensity. Hereby, the modal overlap between the two modes varies from left to right with a maximal overlap in the center of the beam. In addition, the phase of each mode is changed from the top to the bottom of the square shaped beam. For one component the phase increases from 0 to π, for the other mode it decreases from π to 0. To finally create the custom-tailored Poincaré beam, one beam has to be vertically polarized and the other horizontally polarized (following the ideas of the former vector beams). In mathematical terms the obtained Poincaré beam can be written as

$$\mathbf{Vec}^{ctP} = \frac{1}{\sqrt{2}}\Big(\underset{\rightarrow}{u}\,\exp(i\downarrow)\cdot\mathbf{e}_V + e^{i\varphi}\,\underset{\leftarrow}{u}\,\exp(i\uparrow)\cdot\mathbf{e}_H\Big). \tag{2.27}$$

where u stands for the square shaped spatial mode and the arrows depict the direction of increasing intensity (horizontal arrows) or phase (vertical arrows). The resulting Poincaré beam pattern can be seen in Fig. 2.9. Similar to Poincaré modes every polarization can be found in the arising polarization patterns. Hence, the term "custom-tailored Poincaré beams" is justified. The generated pattern changes continuously as

[5]Although the square shaped modes might not be a simple solution to the PWE, they should be describable with a complex superposition of many "simple" modes (maybe an infinite sum of modes). This stems from the fact that all earlier discussed modes form a complete set of spatial modes. Thus, they are able to describe any paraxial light field. A more detailed discussion however is outside of the scope of this thesis.

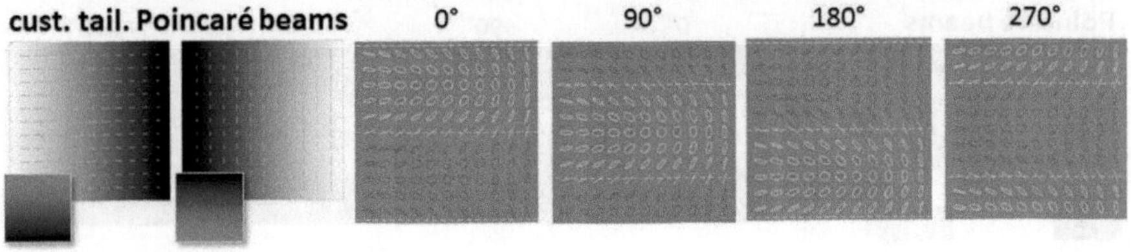

Fig. 2.9 Polarization pattern of a custom-tailored Poincaré beams. Again, the background shows the calculated intensity in *gray* scale. For superpositions (columns three to six) the intensity distribution is constructed to be flat. The polarization pattern is displayed in the foreground (shapes and colors of the polarization vectors depict the type of polarization; color coding as in Fig. 2.6). Again, the first two columns show the custom-tailored modes and their phase distribution (small insets) that are superposed to construct the custom-tailored Poincaré beams. Similar to the Poincaré beams discussed previously, C-points and L-line (*dashed lines*) singularities can be revealed

a function of the vertical position from vertical polarization (left side) to horizontal polarization (right side).

2.3 Photons as Quantum Systems

Photonic quantum information is one of the most popular fields in quantum information science. It has already matured to a stage where complex informational tasks can be implemented with advanced technologies and complex experiments [49–51]. Therefore, the following introduction will be kept as short as possible but as elaborate as necessary to understand the discussed topics in the subsequent chapters. The interested reader is directed to the excellent standard textbooks by Nielsen and Chuang [52], Bouwmeester et al. [53] or Audretsch [54] for further information.

2.3.1 *Quantum States*

So far, light has been described classically by electromagnetic fields. However, in the limit of very weak intensities this wave-like picture in not tenable anymore and light has to be described by single photons and their quantum states (for a detailed discussion about the theoretical justification, I refer to the book [55]). Generally, quantum states are represented by state vectors in a complex Hilbert space. For the sake of simplicity, the following considerations will be restricted to states of the two-dimensional Hilbert space, called "qubits". In this thesis all investigated states are mere qubit states. However, most photonic degrees of freedom can be used to encode quantum states of higher dimensionality of which a few will be named in Sect. 2.3.2.

A general qubit state in the Hilbert space $\mathcal{H}$ is described in Dirac notation by

$$|\psi\rangle = \cos(\theta/2)\,|0\rangle + e^{i\varphi}\sin(\theta/2)\,|1\rangle, \tag{2.28}$$

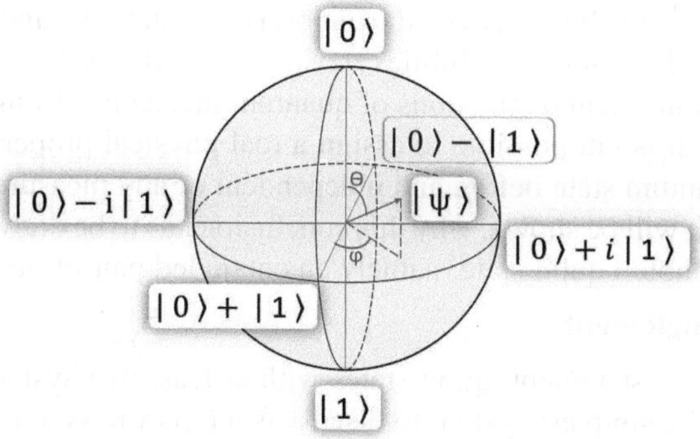

Fig. 2.10 Bloch sphere for qubit systems. Any two-dimensional pure state is related to a specific point on the surface of the Bloch sphere. The state vectors of the computational basis $|0\rangle$ and $|1\rangle$ are found to be on the poles, equally weighted superpositions along the equator. As an example, the state ψ from Eq. (2.28) with the according angles is shown in *red*

where $0 \leq \theta \leq \pi$ and $0 \leq \varphi \leq 2\pi$.

The states $|0\rangle$ and $|1\rangle$ are usually called computational basis and label two mutually exclusive states of the system, i.e. properties of a two-level system described by quantum mechanical states. This means that the physical property under investigation is in a superposition of two different states, each will be detected with a probability of $|\cos(\theta/2)|^2$ for "0" and $|\sin(\theta/2)|^2$ for "1". Besides the computational basis $|0\rangle$ and $|1\rangle$, one can define more, fully equivalent bases to describe the qubit state. Three equivalent bases can be constructed in total such that the measured result in one basis does not give any information about possible results in another basis. Hence, they are mutually unbiased. All three bases together form a set of mutually unbiased bases (MUB). For qubits they can be written as

$$|0\rangle\,, |1\rangle\,; \quad \frac{1}{\sqrt{2}}|0\rangle + |1\rangle\,, \frac{1}{\sqrt{2}}|0\rangle - |1\rangle\,; \quad \frac{1}{\sqrt{2}}|0\rangle + i\,|1\rangle\,, \frac{1}{\sqrt{2}}|0\rangle - i\,|1\rangle\,. \quad (2.29)$$

MUBs play an important role in various quantum informational tasks like the verification of quantum entanglement (as it will shown in Sect. 2.4), optimal state tomography [56], quantum cryptography schemes, and quantum computation [52, 53].

All states governed by Eq. (2.28) can be represented by the surface of a unit sphere, called the Bloch sphere (see Fig. 2.10). Note, that even mixed states are represented as they would lie within the Bloch sphere with the completely mixed state located in the centre.[6] Physical implementations of the states might be (apart from photonic degrees of freedom discussed in the following section) 2-level atoms, ions in two energy levels, the spin of electrons or nucleus, quantum dots, and superconducting currents, to name a few.

---

[6] I abstain from discussing in detail the more realistic description of physical states in terms of density matrices for mixed states and only refer the reader to the corresponding literature [52, 54]. Nevertheless, the simplifying assumption of pure states is a very good approximation to all real physical systems in the presented results. Thus, all drawn conclusions hold.

The aforementioned link between the unobservable state $|\psi\rangle$ and the real observations, namely the detection probabilities $|\cos(\theta/2)|^2$ and $|\sin(\theta/2)|^2$, already shows one of the most important distinctions of quantum mechanics from all classical theories in physics. It is not possible to assign a real physical property to the abstract concept of a quantum state before and independent of any measurement. In the following sections it will be shown, why this conclusion has to be drawn by the example of a specific bi-photon qubit state, namely an entangled pair of qubits.[7]

Quantum Entanglement

To justify the latter statement, qubit states with at least two systems will be investigated. Hence, the simplest system to discuss is a bipartite system.[8] A general two photon qubit-state in the product Hilbert space $\mathcal{H}^{AB} = \mathcal{H}^A \otimes \mathcal{H}^B$ can be written as

$$|\Psi\rangle_{AB} = \sum_{r,s=0}^{1} \lambda_{rs} \, |r\rangle_A \, |s\rangle_B \tag{2.30}$$

$$= \sum_{r,s=0}^{1} \lambda_{rs} \, |rs\rangle_{AB}, \tag{2.31}$$

where the indices label the two photonic systems $|r\rangle_A \in \mathcal{H}^A$ and $|s\rangle_B \in \mathcal{H}^B$ and the amplitudes λ_{rs} are complex numbers that satisfy $\sum \lambda_{rs}^2 = 1$.

The state $|\Psi\rangle$ is called separable if it can be written as a product state

$$|\Psi^{sep}\rangle_{AB} = |\psi\rangle_A \otimes |\psi\rangle_B. \tag{2.32}$$

Here, each subsystem behaves as a fully separable, individual system without any relation to the other subsystem. However, the more interesting case appears when the amplitudes are $\lambda_{rs} = 0$ either for $r = s$ or $r \neq s$. The two possible states

$$|\Phi^{ent}\rangle_{AB} = \lambda_{00} \, |0\rangle_A \, |0\rangle_B + \lambda_{11} \, |1\rangle_A \, |1\rangle_B \tag{2.33}$$

and

$$|\Psi^{ent}\rangle_{AB} = \lambda_{01} \, |0\rangle_A \, |1\rangle_B + \lambda_{10} \, |1\rangle_A \, |0\rangle_B \tag{2.34}$$

are called entangled. They cannot be rewritten as product states. Entangled states describe the bipartite system as a superposition of a single, combined system [59], which leads immediately to surprising consequences. The two entangled subsystems can be physically separated but the observation of one is strongly correlated to the

[7]Note that even single higher-dimensional systems—so-called qudit states—could also be used to justify the statement (for example it was shown for single qutrits in [57]). However, they are not in the scope of this thesis, where qubit systems are the focus of all investigations.

[8]Although multi-partite systems of more than two parties can form very interesting in quantum states, e.g. GHZ states [58], they are not in the scope of this thesis.

measurement result of the other. Although a more detailed discussion will be pursued later in this chapter (Sect. 2.4), it becomes apparent that quantum entanglement is *the* mind boggling feature of quantum physics, which led to a drastic change in the understanding of nature.

Before discussing different ways to encode quantum information and various tests to verify entanglement, a frequently used subset of entangled bipartite qubit systems, the so-called Bell-states, should be introduced:

$$|\Phi^+\rangle_{AB} = \frac{1}{\sqrt{2}}\left(|0\rangle_A\,|0\rangle_B + |1\rangle_A\,|1\rangle_B\right) \tag{2.35}$$

$$|\Phi^-\rangle_{AB} = \frac{1}{\sqrt{2}}\left(|0\rangle_A\,|0\rangle_B - |1\rangle_A\,|1\rangle_B\right) \tag{2.36}$$

$$|\Psi^+\rangle_{AB} = \frac{1}{\sqrt{2}}\left(|0\rangle_A\,|1\rangle_B + |1\rangle_A\,|0\rangle_B\right) \tag{2.37}$$

$$|\Psi^+\rangle_{AB} = \frac{1}{\sqrt{2}}\left(|0\rangle_A\,|1\rangle_B - |1\rangle_A\,|0\rangle_B\right) \tag{2.38}$$

Bell-states are maximally entangled states and form a basis for the bipartite qubit states. They play a special role in many experimental and theoretical investigations. Here, the Bell-state Ψ^- plays a special role because it is the only state that is anti-correlated in all three MUBs. It is often called the qubit singlet state.

2.3.2 Ways to Encode Photonic Quantum Information

In the previous section quantum states have been considered mainly as abstract vectors in a two-dimensional Hilbert space. This section will focus on physical realizations of photonic quantum states.

There are many different ways to encode quantum information on single photons. Theoretically, every photonic degree of freedom (DOF) could be utilized to generate superpositions or entanglement. However, technical difficulties complicate the implementations in quantum optical laboratories. For the presented findings, two DOF are of special importance and will be discussed separately: polarization and transverse spatial modes. For the sake of completeness, other DOF for which photonic entanglement has been demonstrated should also be mentioned: phase and momentum [60], energy and time [61], position and momentum [62] or discrete colors [63]. Most of these systems work with continuous properties, which can be discretized to create not only qubit but higher-dimensional quantum states. Two popular examples are time-bin [64, 65] or path entangled photons [66, 67].

Polarization

Polarization is the most widely used DOF in quantum information science realized with single photons [68]. Ways to create, manipulate and measure the polarization states of single photons like down conversion crystals, wave plates, polarization filters, and polarizing beam splitters are matured technologies. As described in the previous sections for classical fields, polarization is limited to two mutually exclusive states. Therefore, polarization naturally forms a two-dimensional quantum state, which can be represented by the introduced qubit formalism.

The MUBs are represented by the following polarizations

$$\text{horizontal}: |H\rangle = |0\rangle \tag{2.39a}$$

$$\text{vertical}: |V\rangle = |1\rangle \tag{2.39b}$$

$$\text{diagonal}: |D\rangle = \frac{1}{\sqrt{2}}(|H\rangle + |V\rangle) = \frac{1}{\sqrt{2}}(|0\rangle + |1\rangle) \tag{2.39c}$$

$$\text{anti-diagonal}\, |A\rangle = \frac{1}{\sqrt{2}}(|H\rangle - |V\rangle) = \frac{1}{\sqrt{2}}(|0\rangle - |1\rangle) \tag{2.39d}$$

$$\text{right-handed circular}\, |R\rangle = \frac{1}{\sqrt{2}}(|H\rangle + i\,|V\rangle) = \frac{1}{\sqrt{2}}(|0\rangle + i\,|1\rangle) \tag{2.39e}$$

$$\text{left-handed circular}\, |L\rangle = \frac{1}{\sqrt{2}}(|H\rangle - i\,|V\rangle) = \frac{1}{\sqrt{2}}(|0\rangle - i\,|1\rangle) \tag{2.39f}$$

Again, the MUBs define the axes of a three-dimensional sphere, called Poincaré sphere (see Fig. 2.11 on the left). The linear polarization states, which include the computational basis (Eqs. (2.39a) and (2.39b)), are distributed around the equator, while the poles are constituted by right and left-handed circular polarization. Since all MUBs can be used to describe any state equivalently, the difference to the Bloch sphere (Fig. 2.10) is only a matter of convention.

Transverse Spatial Modes

Although spatial modes of light have been known for a long time, their use in quantum physics started quite recently. In the seminal paper from Mair et al. [70] in 2001 the entanglement of LG modes was shown for the first time. Shortly after that paper, an interesting feature of spatial mode entanglement, namely the possibility to create higher-dimensional entanglement, was shown in another experiment [71]. Since then, LG modes have been used in many different quantum experiments for various tasks like quantum bit commitment [72] quantum coin tossing [73], high-dimensional cryptography [74], higher-dimensional entanglement [75] even up to 100 dimensions [76], and novel uncertainty relations in angle and angular momentum [77, 78].

Although the main focus will lie on LG modes, it is worthwhile mentioning, that the entanglement of other modes, like OAM containing Bessel-Gauss modes [79] or Ince-Gauss modes [80], have also been demonstrated recently. Latter ones are especially interesting since they have an additional parameter, the ellipticity. Here,

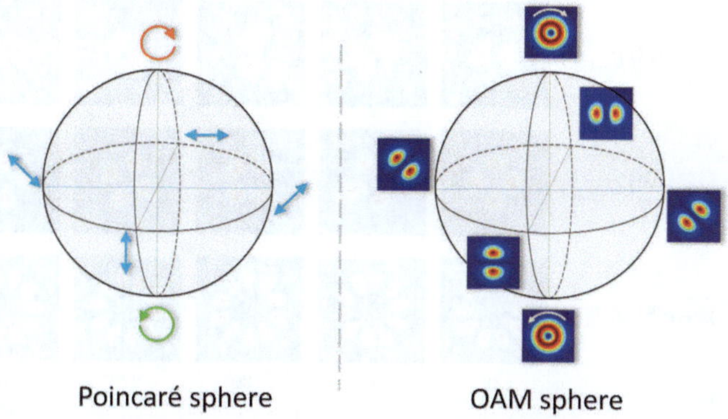

Fig. 2.11 Bloch sphere analogs for polarization and spatial modes. *Left* Poincaré sphere for polarization, where the poles are typically represented by circularly-polarized photons. Along the equator, the linear polarization rotates by 180° through all possible linear polarization states. *Right* OAM sphere analog for a two-dimensional subspace of LG modes (here the first-order modes with $l = \pm 1$). Similar to the linear polarization on the Poincaré sphere, the superposition structure rotates by 180° along the equator. As will be explained later, this angle of rotation is l-times smaller for higher-order LG modes (see Fig. 2.12) [5, 69]

two modes of the same order (same quantum numbers p and m) can differ or even be close to orthogonal (mutually exclusive) if the ellipticity for both modes is different. This IG gauss specific feature might be advantageous as an extension to already known cryptography protocols (for more information see [80]).

The general notation for two-dimensional quantum states encoded in spatial modes will be analogous to the one presented for polarization. The analog to the Poincaré sphere, the two-dimensional OAM sphere for $l = \pm 1$ can be seen in Fig. 2.11. The poles are defined by the LG mode with right-handed or left-handed OAM, respectively, whereas along the equator equally weighted superpositions can be found. These superpositions show an interesting feature, namely they all have the same intensity structure, which rotates with the varying phase of the superposition. Because measurements in two of three MUBs, e.g. measurements around the equator, are sufficient to verify entanglement (which will be discussed in the following section in detail), it is interesting to investigate this feature in more detail. A superposition of LG modes of the same order l can be written as

$$|\psi_{LG}\rangle = \cos(\theta/2)\,|LG_{l,p}\rangle + e^{i\varphi}\sin(\theta/2)\,|LG_{-l,p}\rangle, \tag{2.40}$$

where the phase is φ and the amplitudes are weighted by θ. With equal weighting ($\theta = \pi/2$) and only looking at the angle ϑ dependence of the intensity $I(\vartheta)$, it follows

$$I(\vartheta) \propto \cos^2\left[\left(\frac{\varphi}{2l} - \vartheta\right)\cdot l\right]. \tag{2.41}$$

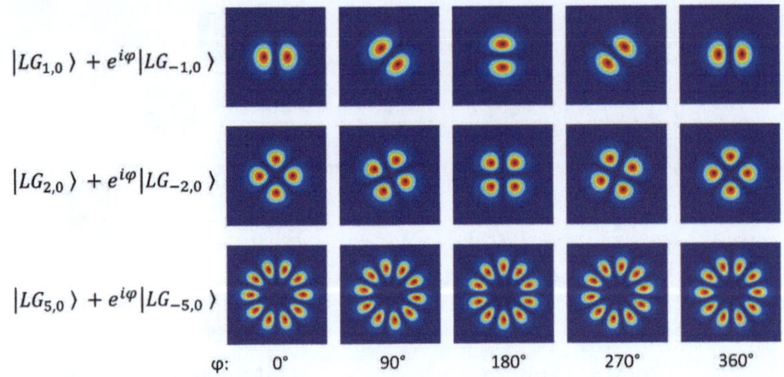

Fig. 2.12 Intensity structures of superpositions for different-order LG beams. The intensity shows $2l$ maxima arranged in a ring. Thet rotation angle γ is depended on the phase φ

There are $2l$ intensity maxima if the angle ϑ is varied from 0 to 2π, which corresponds to a closed loop around the optical axis. Moreover, changing the phase φ of the superposition leads to a simple rotation of the structure, where its angular position γ (in degree) is connected to the phase as follows

$$\gamma = \frac{\varphi}{2l}\frac{360°}{2\pi}. \tag{2.42}$$

The intensities of different superpositions and various phases can be seen in Fig. 2.12. Importantly, it can be seen that the angular position of the intensity for superpositions of $\pm l$ are directly related to the phase of the superposition. Hence, every equally weighted superposition (two of three MUBs) can be accessed by only measuring the intensity structure and its angular position. Before introducing a novel way of measuring LG mode superpositions, which takes advantage of this feature (Chap. 3), it will be shown in the following section that measuring two of three MUBs is enough to prove entanglement.

2.4 Tests of Quantum Entanglement

There are many different methods to show entanglement of the created quantum state. A very powerful way is to identify strict bounds for classical (separable) states, which have to be tested under specific, properly defined assumptions. For the experimental detection of entanglement, this generally leads to an inequality, which can only be violated if the measured data is based on entangled quantum systems.

Three different popular types of such entanglement tests will be briefly discussed in the following sections (for more details see [81]), starting with the one that makes the strongest assumptions, a so-called entanglement witness.

2.4.1 Entanglement Witness

Entanglement witnesses fully rely on quantum mechanical assumptions. The maximum expectation value of the witness for general bipartite separable states, like

$$|\Psi^{sep}\rangle = |\psi_1\rangle \otimes |\psi_2\rangle \tag{2.43}$$

$$= \Big(\cos(\theta_1/2)\,|0\rangle + e^{i\varphi_1}\sin(\theta_1/2)\,|1\rangle\Big) \otimes \Big(\cos(\theta_2/2)\,|0\rangle + e^{i\varphi_2}\sin(\theta_2/2)\,|1\rangle\Big), \tag{2.44}$$

is utilized as an upper bound. If the bound is exceeded, the system is entangled. In the following, two slightly different types of entanglement witnesses that measure the visibilities in the MUBs will be derived.

Witness in 3 MUBs

The first witness operator can be constructed as follows:

$$\hat{W}_3 = \hat{\sigma}_x \otimes \hat{\sigma}_x + \hat{\sigma}_y \otimes \hat{\sigma}_y + \hat{\sigma}_z \otimes \hat{\sigma}_z, \tag{2.45}$$

where $\hat{\sigma}_x$, $\hat{\sigma}_y$ and $\hat{\sigma}_z$ stand for the single-qubit Pauli matrices for the bipartite systems (positions denote the systems).

Measurements of the single-qubit Pauli matrices for a bipartite system correspond to the visibility measurements in the MUB denoted by the index. Thus, the different terms of the witness $\hat{W}_3$ can be rewritten in terms of projection operators

$$\hat{P}_{i,i} = |i,i\rangle\,\langle i,i|, \tag{2.46}$$

where i can stand for any vector of all MUBs (see Eq. (2.29)). This leads e.g. for the x-basis to

$$< \hat{\sigma}_x \otimes \hat{\sigma}_x > = vis_x \tag{2.47}$$

$$= \left| \frac{< \hat{P}_{+,+} > + < \hat{P}_{-,+} > - < \hat{P}_{+,-} > - < \hat{P}_{-,-} >}{< \hat{P}_{+,+} > + < \hat{P}_{-,+} > + < \hat{P}_{+,-} > + < \hat{P}_{-,-} >} \right|, \tag{2.48}$$

where the indices + and − denote "$|0\rangle + |1\rangle$" and "$|0\rangle - |1\rangle$", respectively. In order to be independent from correlated or anti-correlated measurements the absolute value is used. The same holds for the other MUBs y and z too.

The index of $\hat{W}_3$ labels the number of MUBs taken into account. The expectation value for general separable states can be calculated as

$$W_3 := \langle \Psi^{sep}|\,\hat{W}_3\,|\Psi^{sep}\rangle = \sin(\theta_1)\cdot\sin(\theta_2)\cdot\cos(\varphi_1-\varphi_2) + \cos(\theta_1)\cdot\cos(\theta_2), \tag{2.49}$$

which has a maximal value of 1. Hence, the entanglement criterion written in more "experimental" terms of visibilities vis_i is

$$W_3 = vis_x + vis_y + vis_z \quad \begin{cases} \leq 1 & \text{separable} \\ > 1 & \text{entangled.} \end{cases} \tag{2.50}$$

Maximally entangled states have perfect correlations (or anti-correlations) in all three MUBs, i.e. visibilities of 1 for each MUB, which leads to a maximum achievable value of 3 within the quantum mechanical framework. Therefore, the classical bound of 1 can be violated with entangled bipartite systems where an average visibility of >30 % can be measured.

Witness in 2 MUBs

So far, the calculation of the classical bound for measurements in all three possible MUBs was considered. However, sometimes it might be simpler not to measure all of them due to experimental constraints. If only two of the three MUBs are taken into account e.g. σ_x and σ_y (as is it actually the case for some of the presented measurements in this thesis), the witness can be reformulated to

$$\hat{W}_2 = \hat{\sigma}_x \otimes \hat{\sigma}_x + \hat{\sigma}_y \otimes \hat{\sigma}_y. \tag{2.51}$$

A similar straight forward calculation of the expectation value for separable bi-photon states yields

$$W_2 := \langle \Psi^{sep} | \hat{W}_2 | \Psi^{sep} \rangle = \sin(\theta_1) \cdot \sin(\theta_2) \cdot \cos(\varphi_1 - \varphi_2), \tag{2.52}$$

which is also upper bounded by 1. Again, the entanglement criterion for measurements in 2 MUBs can be written in a similar manner as before

$$W_2 = vis_x + vis_y \quad \begin{cases} \leq 1 & \text{separable} \\ > 1 & \text{entangled} \end{cases} \tag{2.53}$$

Because the bound has the same value as before but less measurements are taken into account, the verification of entanglement gets more challenging in the experiment. In other words, the highest obtainable value for maximally entangled states is only 2. Hence, it is harder (the average visibility in both MUBs needs to be >50 %) to prove entanglement by only measuring the visibilities in two MUBs.

2.4.2 Steering-Inequality

A second way to determine if the bipartite system is entangled or not is the test of a steering-inequality. The concept of "steering" as a feature of quantum mechanics was originally discussed by Schrödinger [82]. Although it can be used to test entanglement, i.e. it is a subclass of more general tests for non-separability like entanglement witnesses, it actually makes weaker assumptions in its derivation, thus excludes a broader class of states. In the derivation of the steering-inequality, which was introduced by Wiseman and co-workers [83], only one party requires measurable probabilities in agreement with quantum mechanics (like in the witness derivation), while the other accepts probabilities obeying local realism.[9] Therefore, by demonstrating non-classicality with a steering-inequality not only non-separability is proven but also the non-local steering of the investigated property (the interested reader is directed to [82–85] for more information and the derivation of the inequality).

The steering-inequality is

$$S_{St} = |\hat{\sigma}_x \otimes \hat{\sigma}_x|^2 + |\hat{\sigma}_y \otimes \hat{\sigma}_y|^2 + |\hat{\sigma}_z \otimes \hat{\sigma}_z|^2 \leq 1. \tag{2.54}$$

Compared to the witness criterion (2.50), the steering-inequality is violated by a smaller class of states. Mathematically, this is expressed by squaring of the three terms. Again, the quantum mechanical prediction for maximally entangled states is three. The steering-inequality was already successfully tested in various experiments [86, 87] even in a loophole free manner [88].

2.4.3 Bell-CHSH-Inequality

The oldest and by far most popular way to show entanglement in an experiment, is the test of a Bell-inequality. Nearly twenty years after the first discussion of Einstein et al. [89], Bell discovered a way to experimentally distinguish between the local-realistic description of the world and the one, which quantum mechanics describes [90]. Besides all philosophical implications, this enabled experimentalists to test if entanglement can be found in the investigated system. Bell found an experimentally testable inequality, which is upper bounded by any local realistic theory. A separable two qubit quantum state can be regarded as such a local-realistic model. If the experiment shows a violation of this so-called Bell-inequality, there is no separable description of the state. Thus, the state must be entangled. As seen in the previous sections, nowadays there are different ways to verify entanglement. However testing a Bell-inequality is still the most important proof, since in its derivation nothing but a local-realistic description of the world is assumed (for both parties). Therefore, only the experimental violation of a Bell-inequality strictly proves the non-classicality of nature (assuming all loopholes are closed, which has not yet been achieved in a single experimental test).

[9] Without explaining the concept of local realism in more detail, I use the modern, standard way of defining locality, i.e. two space-like separated systems cannot influence each other faster than the speed of light, and realism, i.e. measurable properties do not exist before and independent of the measurements.

In this thesis the well-known Bell-CHSH-inequality will be used (for its derivation I refer to the original work by Clauser et al. [91]):

$$S = |E(\alpha, \beta) - E(\alpha', \beta) + E(\alpha, \beta') + E(\alpha', \beta')| \leq 2 \tag{2.55}$$

where α, α', β, and β' label different measurement settings and E is the normalized correlation function for photon pairs to be found with these settings. E can be calculated from the measurements as follows:

$$E(\alpha, \beta) = \frac{C(\alpha, \beta) - C(\alpha^{\perp}, \beta) - C(\alpha, \beta^{\perp}) + C(\alpha^{\perp}, \beta^{\perp})}{C(\alpha, \beta) + C(\alpha^{\perp}, \beta) + C(\alpha, \beta^{\perp}) + C(\alpha^{\perp}, \beta^{\perp})} \tag{2.56}$$

where C denotes the measured coincidence count rates at the settings α, β. The superscript $\perp$ indicates the orthogonal setting.

If the measured result surpasses the upper limit of 2, the measured state is entangled, to be more precise, it is not explainable by local realism. Maximally entangled states can reach the quantum mechanical limit of $2\sqrt{2}$.

All three presented types of entanglement tests can be used to reveal entanglement and show a certain hierarchy: Demonstrating entanglement by a Bell-test is harder than by a steering-inequality. Similarly, testing entanglement by a steering-inequality is more challenging than proving it by an entanglement witness. The hierarchy directly follows from the strength of the assumptions that have to be made during the derivation and was proven using Werner states and projective measurements in [83, 84].

With this, the main theoretical considerations, which will be required later, have been completed. In the following, three different experiments that have been conducted during the time of the thesis will be presented and the results will be discussed. At first, a novel flexible way to efficiently create arbitrary spatial mode entanglement will be introduced. It will be used in a first experiment to entangle two photons with very high orbital angular momentum quanta.

References

1. Born, M., & Wolf, E. (1999). *Principles of optics: Electromagnetic theory of propagation, interference and diffraction of light*. CUP Archive.
2. Saleh, B. E. A., Teich, M. C., & Saleh, B. E. (1991). *Fundamentals of photonics* (Vol. 22). New York: Wiley.
3. Siegman, A. E. (1986). *Lasers* (Vol. 37). Mill Valley, CA: University Science Books.
4. Andrews, D. L., & Babiker, M. (2012). *The angular momentum of light*. Cambridge: Cambridge University Press.
5. Mair, A. (2000). *Nichtlokale und Singuläre Quantenzustände des Lichtsd*. Ph.D. thesis, Institut für Experimentalphysik der Universität Wien.
6. Lax, M., Louisell, W. H., & McKnight, W. B. (1975). From maxwell to paraxial wave optics. *Physical Review A, 11*, 1365.
7. Davis, L. W. (1979). Theory of electromagnetic beams. *Physical Review A, 19*, 1177.

8. Bialynicki-Birula, I., & Bialynicka-Birula, Z. (2013). The role of the Riemann-Silberstein vector in classical and quantum theories of electromagnetism. *Journal of Physics A: Mathematical and Theoretical*, *46*, 053001.
9. Moon, P., & Spencer, D. E. (1971). Eleven coordinate systems. *Field theory handbook* (Vol. 1). Berlin: Springer.
10. Krenn, M. (2012). Investigation of complex spatial mode structures of photons. Master's thesis, Technische Universität Wien.
11. Bandres, M. A., Gutiérrez-Vega, J. C., & Chávez-Cerda, S. (2004). Parabolic nondiffracting optical wave fields. *Optics Letters*, *29*, 44.
12. López-Mariscal, C., Bandres, M., Gutiérrez-Vega, J., & Chávez-Cerda, S. (2005). Observation of parabolic nondiffracting optical fields. *Optics Express*, *13*, 2364.
13. Bandres, M. A., & Gutiérrez-Vega, J. C. (2007). Cartesian beams. *Optics Letters*, *32*, 3459.
14. Bandres, M. A., & Gutiérrez-Vega, J. C. (2008). Circular beams. *Optics Letters*, *33*, 177.
15. Bandres, M. A., & Gutiérrez-Vega, J. C. (2008). Elliptical beams. *Optics Express*, *16*, 21087.
16. Bandres, M. A., & Gutiérrez-Vega, J. C. (2004). Ince Gaussian beams. *Optics Letters*, *29*, 144.
17. Pampaloni, F., & Enderlein, J. (2004). Gaussian, Hermite-Gaussian, and Laguerre-Gaussian beams: A primer, arXiv preprint physics/0410021.
18. Langford, N. K. (2007). Encoding, manipulating and measuring quantum information in optics. Ph.D. thesis, University of Queensland.
19. Franke-Arnold, S., Allen, L., & Padgett, M. (2008). Advances in optical angular momentum. *Laser and Photonics Reviews*, *2*, 299.
20. López-Mariscal, C., & Gutiérrez-Vega, J. C. (2007). Propagation dynamics of helical Hermite-Gaussian beams. In *Proceedings of SPIE* (Vol. 6663).
21. Plick, W. N., Krenn, M., Fickler, R., Ramelow, S., & Zeilinger, A. (2013). Quantum orbital angular momentum of elliptically symmetric light. *Physical Review A*, *87*, 033806.
22. Poynting, J. H. (1909). The wave motion of a revolving shaft, and a suggestion as to the angular momentum in a beam of circularly polarised light. *Proceedings of the Royal Society of London A*, *82*, 560.
23. Beth, R. A. (1936). Mechanical detection and measurement of the angular momentum of light. *Physical Review*, *50*, 115.
24. Allen, L., Beijersbergen, M. W., Spreeuw, R. J. C., & Woerdman, J. P. (1992). Orbital angular momentum of light and the transformation of Laguerre-Gaussian laser modes. *Physical Review A*, *45*, 8185.
25. He, H., Heckenberg, N. R., & Rubinsztein-Dunlop, H. (1995). Optical particle trapping with higher-order doughnut beams produced using high efficiency computer generated holograms. *Journal of Modern Optics*, *42*, 217.
26. He, H., Friese, M., Heckenberg, N., & Rubinsztein-Dunlop, H. (1995). Direct observation of transfer of angular momentum to absorptive particles from a laser beam with a phase singularity. *Physical Review Letters*, *75*, 826.
27. O'neil, A., MacVicar, I., Allen, L., & Padgett, M. (2002). Intrinsic and extrinsic nature of the orbital angular momentum of a light beam. *Physical Review Letters*, *88*, 053601.
28. Allen, L., Barnett, S. M., & Padgett, M. J. (2003). *Optical angular momentum*. Boca Raton: Taylor & Francis.
29. Calvo, G. F., Picón, A., & Bagan, E. (2006). Quantum field theory of photons with orbital angular momentum. *Physical Review A*, *73*, 013805.
30. Berry, M. V. (2004). Optical vortices evolving from helicoidal integer and fractional phase steps. *Journal of Optics A: Pure and Applied Optics*, *6*, 259.
31. Leach, J., Yao, E., & Padgett, M. J. (2004). Observation of the vortex structure of a non-integer vortex beam. *New Journal of Physics*, *6*, 71.
32. Zhan, Q. (2009). Cylindrical vector beams: From mathematical concepts to applications. *Advances in Optics and Photonics*, *1*, 1.
33. Maurer, C., Jesacher, A., Fürhapter, S., Bernet, S., & Ritsch-Marte, M. (2007). Tailoring of arbitrary optical vector beams. *New Journal of Physics*, *9*, 78.

34. Galvez, E. J., Khadka, S., Schubert, W. H., & Nomoto, S. (2012). Poincaré-beam patterns produced by nonseparable superpositions of Laguerre-Gauss and polarization modes of light. *Applied Optics*, *51*, 2925.
35. Kano, H., Mizuguchi, S., & Kawata, S. (1998). Excitation of surface-plasmon polaritons by a focused laser beam. *Journal of the Optical Society of America B*, *15*, 1381.
36. Chen, W., & Zhan, Q. (2007). Optimal plasmonic focusing with radial polarization. In *Proceedings of SPIE* (Vol. 6450, p. 64500D).
37. Quabis, S., Dorn, R., Eberler, M., Glöckl, O., & Leuchs, G. (2000). Focusing light to a tighter spot. *Optics Communications*, *179*, 1.
38. Dorn, R., Quabis, S., & Leuchs, G. (2003). Sharper focus for a radially polarized light beam. *Physical Review Letters*, *91*, 233901.
39. Sondermann, M., et al. (2007). Design of a mode converter for efficient light-atom coupling in free space. *Applied Physics B*, *89*, 489.
40. Golla, A., et al. (2012). Generation of a wave packet tailored to efficient free space excitation of a single atom. *The European Physical Journal D*, *66*, 1.
41. Beckley, A. M., Brown, T. G., & Alonso, M. A. (2010). Full poincaré beams. *Optics Express*, *18*, 10777.
42. Nye, J. F. (1983). Lines of circular polarization in electromagnetic wave fields. *Proceedings of the Royal Society of London A*, *389*, 279.
43. Freund, I. (2001). Polarization flowers. *Optics Communications*, *199*, 47.
44. Freund, I., Soskin, M. S., & Mokhun, A. I. (2002). Elliptic critical points in paraxial optical fields. *Optics Communications*, *208*, 223.
45. Dennis, M. R. (2002). Polarization singularities in paraxial vector fields: morphology and statistics. *Optics Communications*, *213*, 201.
46. Soskin, M. S., Denisenko, V., & Freund, I. (2003). Optical polarization singularities and elliptic stationary points. *Optics Letters*, *28*, 1475.
47. Cardano, F., Karimi, E., Marrucci, L., de Lisio, C., & Santamato, E. (2013). Generation and dynamics of optical beams with polarization singularities. *Optics Express*, *21*, 8815.
48. Berry, M. V., & Hannay, J. H. (1977). Umbilic points on Gaussian random surfaces. *Journal of Physics A: Mathematical and General*, *10*, 1809.
49. O'Brien, J. L. (2007). Optical quantum computing. *Science*, *318*, 1567.
50. O'Brien, J. L., & Akira Furusawa, J. V. (2009). Photonic quantum technologies. *Nature Photonics*, *3*, 687.
51. Aspuru-Guzik, A., & Walther, P. (2012). Photonic quantum simulators. *Nature Physics*, *8*, 285.
52. Nielsen, M. A., & Chuang, I. L. (2010). *Quantum computation and quantum information*. Cambridge: Cambridge University Press.
53. Bouwmeester, D., Ekert, A. K., & Zeilinger, A. (2000). *The physics of quantum information* (Vol. 38). Berlin: Springer.
54. Audretsch, J. (2008). *Entangled systems*. Wiley.com.
55. Scully, M. O., & Zubairy, S. (1997). *Quantum optics*. Cambridge: Cambridge University Press.
56. Altepeter, J. B., Jeffrey, E. R., & Kwiat, P. G. (2005). Photonic state tomography. *Advances in Atomic, Molecular, and Optical Physics*, *52*, 105.
57. Lapkiewicz, R., et al. (2011). Experimental non-classicality of an indivisible quantum system. *Nature*, *474*, 490.
58. Greenberger, D. M., Horne, M. A., & Zeilinger, A. (1989). Going beyond bell's theorem. *Bell's theorem, quantum theory and conceptions of the universe* (Vol. 69). Berlin: Springer.
59. Horodecki, R., Horodecki, P., Horodecki, M., & Horodecki, K. (2009). Quantum entanglement. *Reviews of Modern Physics*, *81*, 865.
60. Rarity, J. G., & Tapster, P. R. (1990). Experimental violation of bellś inequality based on phase and momentum. *Physical Review Letters*, *64*, 2495.
61. Kwiat, P. G., Steinberg, A. M., & Chiao, R. Y. (1993). High-visibility interference in a bell-inequality experiment for energy and time. *Physical Review A*, *47*, R2472.
62. Howell, J. C., Bennink, R. S., Bentley, S. J., & Boyd, R. W. (2004). Realization of the Einstein-Podolsky-Rosen paradox using momentum-and position-entangled photons from spontaneous parametric down conversion. *Physical Review Letters*, *92*, 210403.

63. Ramelow, S., Ratschbacher, L., Fedrizzi, A., Langford, N. K., & Zeilinger, A. (2009). Discrete tunable color entanglement. *Physical Review Letters, 103*, 253601.
64. Brendel, J., Gisin, N., Tittel, W., & Zbinden, H. (1999). Pulsed energy-time entangled twin-photon source for quantum communication. *Physical Review Letters, 82*, 2594.
65. de Riedmatten, H., et al. (2004). Tailoring photonic entanglement in high-dimensional Hilbert spaces. *Physical Review A, 69*, 050304.
66. Rossi, A., Vallone, G., Chiuri, A., De Martini, F., & Mataloni, P. (2009). Multipath entanglement of two photons. *Physical Review Letters, 102*, 153902.
67. Schaeff, C., et al. (2012). Scalable fiber integrated source for higher-dimensional path-entangled photonic qunits. *Optics Express, 20*, 16145.
68. Zeilinger, A., Weihs, G., Jennewein, T., & Aspelmeyer, M. (2005). Happy centenary, photon. *Nature, 433*, 230.
69. Padgett, M. J., & Courtial, J. (1999). Poincaré-sphere equivalent for light beams containing orbital angular momentum. *Optics Letters, 24*, 430.
70. Mair, A., Vaziri, A., Weihs, G., & Zeilinger, A. (2001). Entanglement of the orbital angular momentum states of photons. *Nature, 412*, 313.
71. Vaziri, A., Weihs, G., & Zeilinger, A. (2002). Experimental two-photon, three-dimensional entanglement for quantum communication. *Physical Review Letters, 89*, 240401.
72. Langford, N. K., et al. (2004). Measuring entangled qutrits and their use for quantum bit commitment. *Physical Review Letters, 93*, 053601.
73. Molina-Terriza, G., Vaziri, A., Ursin, R., & Zeilinger, A. (2005). Experimental quantum coin tossing. *Physical Review Letters, 94*, 040501.
74. Gröblacher, S., Jennewein, T., Vaziri, A., Weihs, G., & Zeilinger, A. (2006). Experimental quantum cryptography with qutrits. *New Journal of Physics, 8*, 75.
75. Dada, A. C., Leach, J., Buller, G. S., Padgett, M. J., & Andersson, E. (2011). Experimental high-dimensional two-photon entanglement and violations of generalized bell inequalities. *Nature Physics, 7*, 677.
76. Krenn, M., et al. (2014). Generation and confirmation of a (100×100)-dimensional entangled quantum system. *Proceedings of the National Academy of Sciences, 11*, 6243.
77. Jack, B., et al. (2011). Demonstration of the angular uncertainty principle for single photons. *Journal of Optics, 13*, 064017.
78. Leach, J., et al. (2010). Quantum correlations in optical angle-orbital angular momentum variables. *Science, 329*, 662.
79. McLaren, M., et al. (2012). Entangled Bessel-Gaussian beams. *Optics Express, 20*, 23589.
80. Krenn, M., et al. (2013). Entangled singularity patterns of photons in Ince-Gauss modes. *Physical Review A, 87*, 012326.
81. Gühne, O., & Tóth, G. (2009). Entanglement detection. *Physics Reports, 474*, 1–75.
82. Schrödinger, E. (1935). Discussion of probability relations between separated systems. *Proceedings of the Cambridge Philosophical Society, 31*, 555.
83. Wiseman, H. M., Jones, S. J., & Doherty, A. C. (2007). Steering, entanglement, nonlocality, and the Einstein-Podolsky-Rosen paradox. *Physical Review Letters, 98*, 140402.
84. Jones, S. J., Wiseman, H. M., & Doherty, A. C. (2007). Entanglement, Einstein-Podolsky-Rosen correlations, bell nonlocality, and steering. *Physical Review A, 76*, 052116.
85. Cavalcanti, E. G., Jones, S. J., Wiseman, H. M., & Reid, M. (2009). Experimental criteria for steering and the Einstein-Podolsky-Rosen paradox. *Physical Review A, 80*, 032112.
86. Saunders, D. J., Jones, S. J., Wiseman, H. M., & Pryde, G. J. (2010). Experimental EPR-steering using bell-local states. *Nature Physics, 6*, 845.
87. Smith, D. H., et al. (2012). Conclusive quantum steering with superconducting transition-edge sensors. *Nature Communications, 3*, 625.
88. Wittmann, B., et al. (2012). Loophole-free Einstein-Podolsky-Rosen experiment via quantum steering. *New Journal of Physics, 14*, 053030.
89. Einstein, A., Podolsky, B., & Rosen, N. (1935). Can quantum-mechanical description of physical reality be considered complete? *Physical Review, 47*, 777.
90. Bell, J. S. (1964). On the Einstein-Podolsky-Rosen paradox. *Physics, 1*, 195.
91. Clauser, J. F., Horne, M. A., Shimony, A., & Holt, R. A. (1969). Proposed experiment to test local hidden-variable theories. *Physical Review Letters, 23*, 880.

Chapter 3
Entanglement of High Angular Momenta

3.1 Preamble

The content of this chapter is mainly based on the publication "Quantum entanglement of high angular momenta" [1].

The chapter will describe a novel way of entangling two photons in the transverse spatial degree of freedom (DOF) in a very efficient, clean and flexible way. Moreover, it allows the entanglement of very higher-order modes, which carry a large amount of orbital angular momentum (OAM) quanta. To verify the generated entanglement of high OAM, a novel technique was developed to test the non-classicality of the state only by looking at its intensity structure. Both, the novel way to entangle two photons in the transverse spatial DOF and the new method for verifying entanglement, led to the demonstration of entanglement of the highest quantum number that has been experimentally shown till date.

3.1.1 Liquid Crystal Spatial Light Modulator

Before investigating the scheme to entangle photons in their spatial mode, a device that is able to generate such spatial modes is introduced, namely, a liquid crystal spatial light modulators (herein simply called SLM). Because an SLM is able to modulate the phase front of light with very high precision, it has become a standard tool to create or analyze a huge variety of different spatial modes. Therefore, only a brief explanation of the working principle will be given (more information can be found in [2]), followed by the specifications of the type of SLM used in all presented experiments.

SLMs are miniaturized displays consisting of many μm-sized pixels, which can be used individually to modulate the light. The devices used in all following experiments only modulate the phase of the light, leaving the amplitude and polariza-

R. Fickler, *Quantum Entanglement of Complex Structures of Photons*,
Springer Theses, DOI 10.1007/978-3-319-22231-8_3

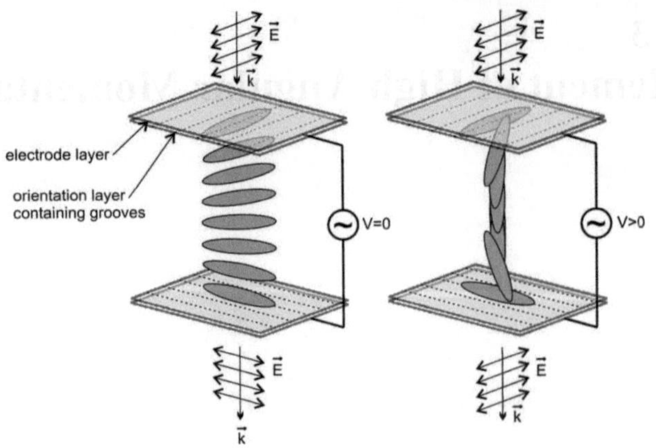

Fig. 3.1 Working principle of a twisted nematic liquid crystal SLM. The liquid crystal is sandwiched between two glass plates with electrodes and alignment grooves. The two parallel glasses are rotated by 90° with respect to each other leading to the twisted configuration. *Left* No voltage is applied, hence the crystals are aligned in a helix-like structure, which leads to a phase modulation of crossing light. *Right* If voltage is applied the crystals align along the electric field and no phase modulation is performed. Pictures are taken from [2]

tion unchanged (latter down to less than 1 %). The basic working principle can be described as follows:

The birefringent liquid crystal (LC, rod-shaped) is in the nematic mesophase (all molecules are oriented the same but randomly distributed) and sandwiched between two glass substrates. On the glass surface, there are transparent electrodes and orientation grooves to force the crystal rods to be aligned along these grooves. The two glasses are rotated by 90° resulting in a twisted orientation of the molecules (see Fig. 3.1). If voltage is applied to the electrodes, the crystals align along the direction of the electric field. Due to the birefringence of the liquid crystal, the incoming light changes its polarization while passing through the crystal. The device is produced such that for a specific polarization, it is the same after leaving as it was before entering the device. However, the polarization change inside the LC leads to dynamic phase shift (optical path difference) plus an additional geometric phase (Pancharatnam-Berry-phase due to polarization change). The SLMs used in all experiments are twisted nematic LC-SLMs and can modulate the phase of the light up to more than 2π. Although the light was described as being transmitted through the SLM, most of the devices available today work in reflection. The light is reflected from a high reflectance mirror after passing through the LC the first time and then passes through the LC again. A sketch of the construction can be seen in Fig. 3.2.

The SLMs that were used are from Holoeye (Fig. 3.2) with a resolution of 1080×1920 pixels (full HD). Each pixel is 8×8 μm in size and can be addressed in up to 256 different values of phase shift. The refresh rate of the display is 60 Hz. The SLM is addressed with a normal computer via the screen port DVI.

Fig. 3.2 *Left* Photo of the utilized device: Holoeye, PLUTO—phase only SLM. *Right* Sketch of SLM display assembly. Both images are taken from the Holoeye web page (www.holoeye.com)

The main advantage of this SLM and the reason why it was used in later experiments is that it has highest resolution commercially available on the market. Therefore, more complex phase patterns can be imprinted onto the beam, which will be important for very high-order LG modes. However, the high resolution comes at a certain cost, namely phase stability of the SLM. Each pixel needs to be addressable with a different voltage, which is hardly feasible with single DC voltage supplies. Therefore, the SLMs from Holoeye are addressed via a so-called pulse-width modulation (PWM) method. Here, an effective DC voltage value is achieved by sending fast pulses with an adjustable width. However, this leads to a small modulation of the molecules inside the LC and thus to flickering of the phase. Although measurable and visible if the modulated light is imaged with a CCD camera, there was no effect detectable in all quantum experiments. A second disadvantage of the SLM from Holoeye stems from its comparable low efficiency. Although up to 80 % of the light was measured to be modulated, only 60 % got reflected (if switched off), which made a total efficiency of around 50 %. All described disadvantages do not exist for example for SLMs from Hamamatsu, but they have less resolution (600×800 pixels).

Adjusting the SLM—Holoeye Pluto

The SLM can modulate the phase of light at 810 nm by more than 2π. However, the modulation is not linear and a modulation more than 2π only reduces the resolution in phase steps due its redundancy. In addition, using the SLM with an incident beam that is not perpendicular to the SLM surface has slight influence on the phase modulation as well. All effects can be corrected in the SLM control unit by adjusting the voltage range of the pulses and adding a lookup table (LUT) for voltage to phase modulation correspondence. To find out the right parameter, a characterization setup was built (see Fig. 3.3). Here, a laser beam is split into two small beams. Both are incident on different regions of the same SLM, thus each beam can be modulated separately by the same SLM. After the beams are reflected from the SLM surface, they are focused by a lens and overlapped in the focal plane. The resulting interference pattern is imaged by a CCD camera. Additionally, a program was developed that tracks the

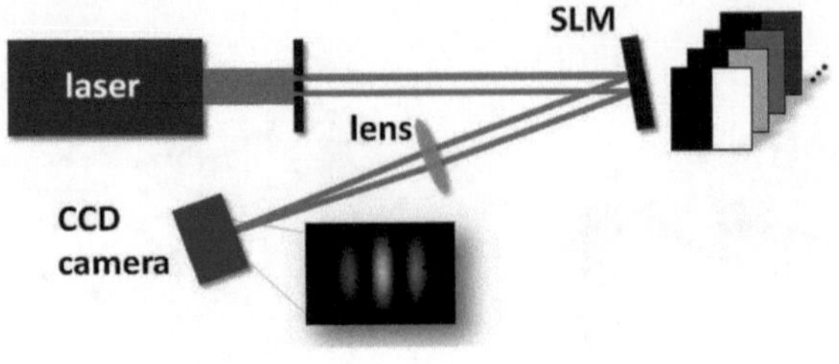

Fig. 3.3 Setup to adjust the phase modulation characteristics. Each beam can be modulated independently with a phase shift corresponding to a *gray* value displayed on the SLM. The interference structure is laterally shifted depending on the relative phase shift between the two beams. The structure is imaged on the CCD camera and tracked by a computer program

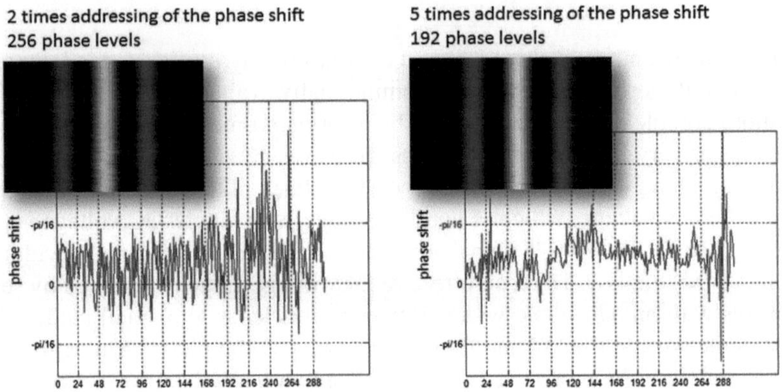

Fig. 3.4 Phase stability of the SLM. *Left* Stability of the default mode, where in each refresh cycle of the SLM (60 Hz) the phase is addressed 2 times and 256 different phase levels are available. *Right* Phase Stability of SLM when the phase shift is addressed 5 times in each frame of the 60 Hz with the cost of "only" 196 different phase levels. The latter mode was used for all experiments

movement of the imaged structure and relates it to the phase modulation of the SLM. Snapshots of the interference structure are taken. From each snapshot, one pixel line is cut out and attached to the line before (see results in Figs. 3.4 and 3.5). A Gaussian distribution is fitted to one maximum of the structure and its position is tracked through the duration of the whole measurement. The distance between two neighboring maxima corresponds two a full 2π phase shift. Thus, any movement of the structure can be related to a phase change of the SLM, either to test the stability or the controlled phase shift.

Phase Stability of the SLM

If the SLM modulates one of the beams by a constant phase shift while the other beam stays unmodulated, the movement of the interference pattern shows the phase stability of the SLM (mechanical stability assumed). Here, different modes of refreshing the liquid crystal display can be tested. Three options are available, where each pixel can

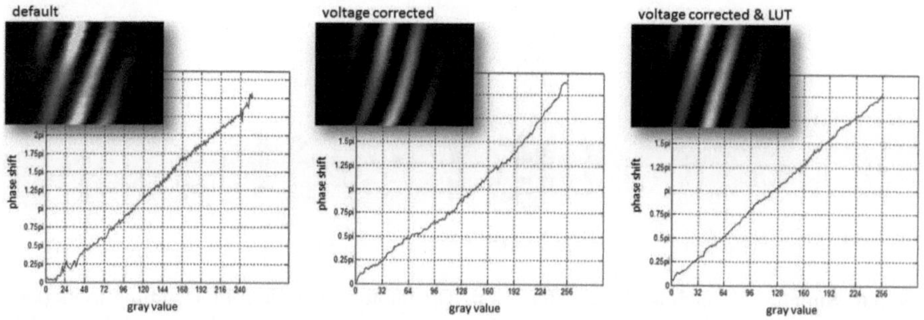

Fig. 3.5 Correction of the phase shift of the SLM. *Left* Uncorrected phase shift of the SLM (default). *Center* The voltage of the PWM is adjusted to reduce the maximum phase shift to approximately 2π. *Right* The final adjustment of the SLM using the phase shift characteristic from the measurement before (*center*) and calculating the compensating look-up-table. The result of the fully corrected SLM phase modulation is shown

either be addressed two, five or eight times in one refresh cycle of the SLM screen (60 Hz). The more often the pixels are refreshed the more stable the phase shift but the less number of addressable phase levels are available (256, 192 or 64 different levels). The second mode where each phase shift is addressed 5 times in one refresh cycle and 192 different phase levels are available was found to be the best trade-off between stability and number of phase levels (see Fig. 3.4).

Linearization of the Phase Shift

If one of the two beams is continuously modulated by the SLM and the movement of the interference pattern is tracked, the phase modulation characteristics can be tested and adjusted. First, the phase is continuously changed by displaying all 256 different gray colors for one beam on the SLM and measure the shift of the interference pattern (see Fig. 3.5 center). A modulation of more than 2π was found and adjusted by lowering the voltage values of the PWM method. Afterwards, a LUT is generated and programmed in the control unit to linearize the modulation of the SLM (see Fig. 3.5 right). An additional advantage of the procedure is that it can be done with the incident angles of the laser beam matching with the later setup layout. Thus, the SLM is perfectly adjusted to the actual experimental conditions.

3.2 Coherent Transfer of Polarization to Transverse Spatial Modes

After the discussion of the characteristics of the SLM, the coherent transfer of the polarization DOF to the transverse spatial modes will be described. The transfer is realized by an interferometric structure (see Fig. 3.6). Once duplicated, it will be used to transfer polarization entanglement to spatial mode entanglement. However now,

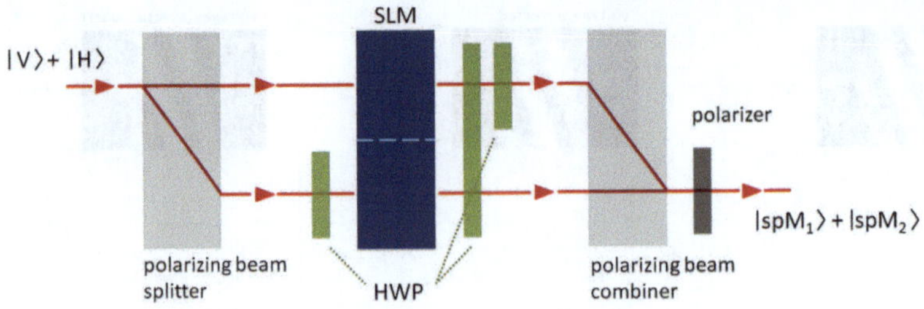

Fig. 3.6 Simplified sketch of the transfer setup. The photon's path is split according to its polarization and transformed to a chosen transverse spatial mode by a spatial light modulator (SLM) depended on the path. Half-wave plates (HWP) were installed into the paths to rotate the polarization such that the SLM works best and the output is separated from the input (see Fig. 3.7). After recombining the paths, a polarizer at 45° projects the photon to diagonal polarization, thereby completing the transfer

only one setup for single photons is considered. The initial state $|\psi\rangle_i$ of one photon that is in a Gauss mode G and in a polarization superposition can be described by

$$|\psi\rangle_i = \Big(\alpha\,|H\rangle + e^{i\varphi}\beta\,|V\rangle\Big) \otimes |G\rangle, \tag{3.1}$$

with $\alpha, \beta \in \mathbb{R}$ and $\alpha^2+\beta^2 = 1$. The path of the photons is split by a polarizing beam splitter into two spatially separated paths. Each path is incident on a different region of the SLM, allowing for separate modulation of the transverse phase profiles dependent on the path. In one path, a half-wave plate (HWP) rotates the polarization before the modulation of the SLM, because the LC display requires a specific polarization to work. After the phase modulation, a second HWP rotates the other polarization such that both are orthogonal again. An additional, third HWP that rotates the polarization in both paths is implemented in order to separate the incoming beam from the outgoing one (see Fig. 3.7). Eventually, the paths are recombined by a polarizing beam combiner. The temporary state $|\psi\rangle_t$ can be written as

$$|\psi\rangle_t = \alpha\,|H, spM_1\rangle + e^{i\varphi}\beta\,|V, spM_2\rangle, \tag{3.2}$$

where spM_1 and spM_2 label the generated spatial modes coupled to the polarization. The transfer has not been completed yet and the state is now in a so-called hybrid superposition of polarization and spatial modes.[1] After recombining the paths, a polarizer projects the photons onto the diagonal basis $|D\rangle = |H\rangle + |V\rangle$, thus erasing any information about which path the photon took and to which spatial mode it was transferred. Hence, the final state $|\psi\rangle_f$ can be written as

$$|\psi\rangle_f = |D\rangle \otimes \Big(\alpha\,|spM_1\rangle + e^{i\varphi}\beta\,|spM_2\rangle\Big). \tag{3.3}$$

[1] As it was shown earlier, the photon state can be considered to be a vector polarization state with transverse spatially varying polarization. See Sect. 2.2.5 for more information.

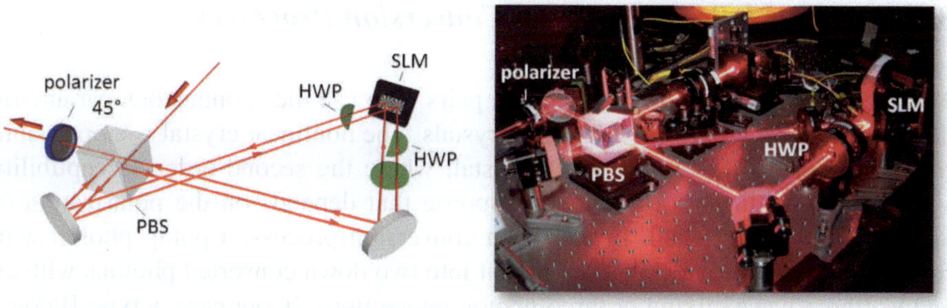

Fig. 3.7 Folded interferometric transfer setup adjusted to experimental requirements. *Left* Sketch of the folded interferometric structure. The paths are split and recombined by the same polarizing beam splitter (PBS). The HWP in the interferometer ensure the right polarization for the SLM and changes it such that the incoming and outgoing paths are separated. *Right* Photo of the transfer setup. Paths are made visible by long exposure time and retracing of the laser beam (photo done by Robin Riegler)

The superposition in polarization has now been transferred coherently to a spatial mode superposition adjustable with the SLM. Therefore, the generation of spatial mode superposition with this transfer setup is neither limited to a specific mode family nor to certain mode order. For the use in real experiments the presented idea has to be adjusted to experimental requirements. Firstly, the SLM reflects the light instead of transmitting it. Secondly, the phase stability of the setup is crucial. Both requirements can be taken into account by realizing the interferometric setup in a folded, Sagnac-like layout (see Fig. 3.7). To maximize the modal overlap after recombination, the layout was balanced such that both paths are of equal length after the modes are generated.

Before the use of the transfer setup (or to be more precise, two transfer setups) in generating any spatial mode entanglement will be discussed, an efficient source for polarization entangled bi-photon states will be explained.

3.3 Source of Polarization Entanglement

It has become a standard procedure in quantum optics laboratories to create photon pairs entangled in polarization from a down-conversion process in a nonlinear crystal. The theory is well understood and many different layouts including different nonlinear crystals have been developed. Hence, the main focus of this section will be a brief review of the type of crystal and entanglement source that are used in later experiments, namely a Sagnac configureration with type-II down-conversion process in a periodically poled KTP crystal (ppKTP). The main reason for the use of that type of entanglement source is its high entangled-pair generation rate. For more details about the source, the reader my be referred to the PhD-thesis of Fedrizzi [3], which was the basis for the presented source.

3.3.1 Photon Pairs from Down Conversion Processes

The most popular way to create photon pairs is to use the spontaneous parametric down conversion process in nonlinear crystals. The nonlinear crystal is a potassium titanyl phosphate $KTiOPO_4$ (KTP) crystal, where the second-order susceptibility $\chi^{(2)}$ is large leading to a nonlinear response that depends on the polarization of an incoming electric field. In the down conversion process, a pump photon with frequency ω_p and wave vector $\mathbf{k}_p$ is split into two down converted photons with ω_i and $\mathbf{k}_i$ (three-wave mixing or three-photon interaction). In our case, a type-II down conversion was utilized where the photons are orthogonally polarized. Conservation of energy and momentum leads to the conditions

$$\omega_p = \omega_1 + \omega_2 \quad ; \quad \mathbf{k}_p = \mathbf{k}_1 + \mathbf{k}_2, \tag{3.4}$$

which are called phase-matching. Again, there are already many different techniques to fulfill the phase-matching conditions. In the presented experiment a periodically poled crystal was used. Here, the effective nonlinearity of the crystal is periodically inverted (by electric fields during the process of crystal growth), which leads to an additional term and the quasi-phase-matching condition

$$\mathbf{k}_p(\lambda_p, n_p(\lambda_p, T)) = \mathbf{k}_1(\lambda_1, n_1(\lambda_1, T)) + \mathbf{k}_2(\lambda_2, n_2(\lambda_2, T)) + \frac{2\pi}{\Lambda(T)}, \tag{3.5}$$

where the wave vectors $\mathbf{k}_i$ are dependent on the wavelength λ_i as well as the wavelength and temperature-depended refractive indices $n_i(\lambda_i, T)$. The poling period Λ is dependent on the temperature T, which allows for simple fine-tuning of the phase-matching conditions in the later experiment. The big advantage of the periodic poling is the great flexibility in custom-tailoring the processes for specific wavelengths and beam angles. The crystal utilized in all experiments was constructed for co-linear down conversion (negligible transversal walk-off) and frequency degenerated photon pairs at 810 nm (pumped with a 405 nm laser).[2]

3.3.2 Sagnac-Type Source of Polarization Entanglement

The nonlinear crystal was placed in the loop of a Sagnac interferometer to not only generate photon pairs, but also polarization entanglement [4, 5]. The idea is fairly simple (see Fig. 3.8). The pump laser enters the Sagnac interferometer in an equally weighted superposition between horizontal and vertical polarization. It is split by a

[2] A measurement of the actual pump and down conversion wavelength identified both values to be 405.5 nm and 811 nm (at 30 °C crystal temperature). However, in all later descriptions the wavelengths will be labeled with "405 nm" for the pump and "810 nm" for the down converted photon pairs.

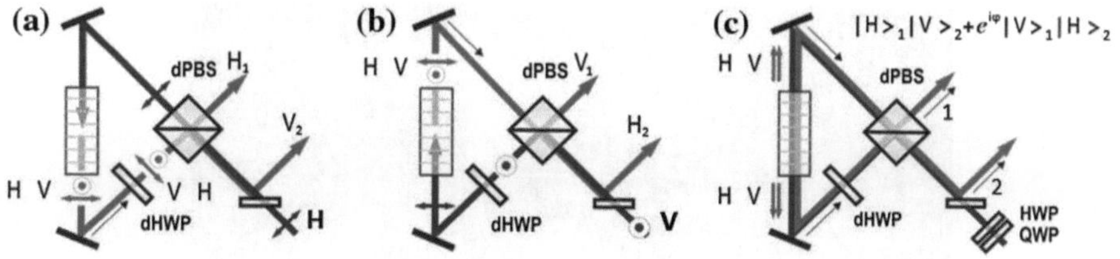

Fig. 3.8 Working principle of a Sagnac-type source for polarization entanglement. **a** Pumping the interferometric setup with a horizontally-polarized beam leads H and V polarized down-converted pairs. **b** Pumping the Sagnac loop with a vertically-polarized beam generates pairs of opposite (compared to case **a**) polarization V and H. **c** Pumping with a coherent equal superposition of horizontal and vertical polarization a polarization entangled state is generated. The phase φ can be adjusted by the pump polarization. (pictures adapted from [3])

dual-wavelength polarization beam splitter (dPBS) working for both wavelengths λ_p and $\lambda_{1,2}$ and pumps the down-conversion crystal in both directions, anti-clockwise and clockwise. An additional half-wave plate (also working for both wavelengths λ_p and $\lambda_{1,2}$) is placed in the interferometer to rotate the polarizations to the required angle for the clockwise pump beam and to pre-compensate the transversal walk-off introduced by the dPBS. The photon pairs are spatially separated by the dPBS and a dichroic mirror (reflective at 405 nm, transmissive at 810 nm). If the pump photon goes anti-clockwise through the interferometer (H polarized), the spatially-separated, down-conversion photon pair will be horizontally (photon 1) and vertically (photon 2) polarized (see Fig. 3.8a). If the pump photon is V polarized (clockwise pumping), the photon pairs would have the orthogonal polarizations, namely vertical (photon 1) and horizontal (photon 2) polarization (see Fig. 3.8b). With the pump beam in equal superposition of H and V, the created bi-photon state $|\Psi\rangle^p$ is entangled in polarization and can be written as

$$|\Psi\rangle^p = \frac{1}{\sqrt{2}}\Big(|H\rangle_1\,|V\rangle_2 + e^{i\varphi}\,|V\rangle_1\,|H\rangle_2\Big), \tag{3.6}$$

where φ is the same phase as the one of the pump beam superposition. Thus, it can be adjusted by simply changing the pump polarization. Analogously, both amplitudes of the pump beam superposition are related to the amplitudes of the entangled pair (for a more detailed discussion see [3]). An unbalanced superposition would lead to a non-maximally entangled state, however, it was not required in any following experiment.

3.3.3 *Characterization of the Polarization Entanglement*

The Sagnac source features a 15 mm long ppKTP crystal with 2 mm width and 1 mm height (photo of the setup in Fig. 3.9). The pump beam power was adjustable up to approximately 32 mW measured before entering the Sagnac interferometer. In front

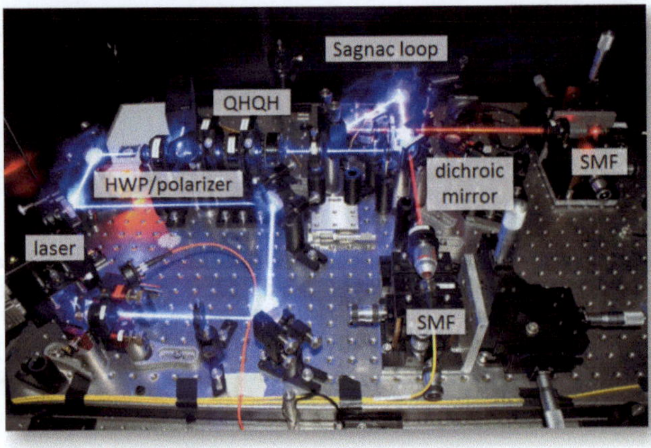

Fig. 3.9 Long exposure time photo from the Sagnac source. Lasers are used to retrace the beam path of the pump and the down converted photons. A half-wave plate (HWP) and a polarizer are used to adjust the pump intensity. A combination of two quarter-wave plates and two half-wave plates (QHQH) is implemented to adjust the polarization of the pump beam. The down converted photons are spatially separated and coupled into single-mode fibers (SMF)

of the single mode fibers, narrow bandpass filters (3 nm) are installed to reduce background noise as well as to enhance the spectral overlap and thereby the quality of entanglement. After coupling into single mode fibers, around 1.2 million entangled photon pairs per second are detected. The average visibility in all three MUBs of entanglement is around 98 % ±0.5 %. The Sagnac source was the basis for all later experiments in which various entanglement witnesses (see Sect. 2.4.1 in the last chapter) are measured. Therefore, the same entanglement tests are conducted for polarization entanglement in the following:
Witness in 3 MUBs, classically bounded by ≤ 1 (Eq. 2.50):

$$2.9340 \pm 0.0006 \tag{3.7}$$

Witness in 2 MUBs, classically bounded by ≤ 1 (Eq. 2.53):

$$1.9445 \pm 0.0006 \tag{3.8}$$

Each of the witnesses verifies entanglement by more than 1500 standard deviations, which demonstrates the quality of the generated polarization entanglement.

After having presented a source for a large number of polarization-entangled photon pairs and a transfer setup to map polarization to any required spatial mode (last section), the combination of the setups can be considered. Two transfer setups coupled to the Sagnac source offer a very flexible way to generate any type of spatial mode entanglement. Similar schemes to entangle different degrees of freedom were discussed in [6] for linear momentum, in [7] for discrete colors and in [8, 9] for transverse spatial modes.

3.4 Creation of High Angular Momenta Entanglement

As mentioned earlier, the transfer setup presented in Sect. 3.2 works for one photon only. If both entangled photons have to be transferred, a duplicate version of the setup needs to be built. Before a detailed description of the whole setup is given, it is briefly explained why it is interesting to entangle high angular momentum carrying modes, followed by an investigation of the limitations of other schemes.

3.4.1 Prospects of High Angular Momenta Entanglement

The motivation for high OAM entanglement is twofold: a fundamental interest in entanglement of high quantum numbers and various aspects for possible applications.

From a fundamental point of view, higher-order LG modes and the OAM associated with them offer an exciting possibility to test the limits of quantum mechanics. Theoretically no upper limit of OAM quanta per a single photon is known, or in other words, how strong the OAM per single photon can be. Moreover, if high OAM values per single photon can be realized, it is interesting to test if quantum features like entanglement can be maintained. Because macroscopic values[3] of OAM should be possible, the obvious question about the general possibility of observing macroscopic quantum effects arises.

On the other hand, possible applications of the entanglement of high OAM quanta give more reasons to increase the OAM quantum number. Firstly, each photon can transfer the high momenta to other systems either by an absorption or reflection process. This might be crucial for example in future opto-mechanical experiments, which currently already employ linear momentum [11]. Moreover, the complex structure of higher-order LG mode superpositions can be used to increase the resolution in angular-position measurements. If the photons are entangled, this enhanced angular sensitivity can be exploited even remotely (see Sect. 3.5.3).

Both branches of reasoning show interesting aspects of entanglement of high angular momentum. The subsequent, natural question to ask is: what are the limits of previous experiments?

3.4.2 Limitations of Previous Experiments

So far, most of the experimental setups generating and testing OAM entanglement go back to the seminal work of Mair et al. [12]. The idea is to use the inherent momentum conservation in the down conversion process to create entanglement in the OAM

[3]One should be careful with the term "macroscopic". There is no generally accepted definition about when an object or a property connected to it, can be considered macroscopic. A possible and loose definition can be found in the paper of Leggett [10].

degree of freedom. If one photon is measured to have $+l$ quanta of OAM, the other will have $-l$ quanta due to angular momentum conservation. The disadvantage of such a scheme is that the generation efficiency of the higher-order modes is directly connected to pump beam dimensions, the length of the crystal and the focusing parameters [13]. In experiments, values up to 20–30 $\hbar$ might be achievable [14–16], however the highest OAM entanglement tested in a recent experiment was 11 $\hbar$ [17].[4] Nevertheless, a recent theoretical idea to increase the possible OAM values in the down conversion by a chirped quasi-phase-matched crystal suggests a promising enhancement [18]. The estimated increase could be tenfold, which would lead to OAM quanta of approx 200–300 $\hbar$ after the down conversion. Although the predicted results are very promising they still rely on the phase matching conditions of the down conversion crystal. On the contrary to these theoretical studies, the presented scheme (source and transfer setups) is only limited by the devices available to generate high-order OAM modes. It can always be adapted to novel possibilities in creating spatial modes containing high quanta of OAM and for testing the limits of quantum mechanics (see e.g. spiral phase mirrors in [19, 20]).

3.4.3 Setup of High Angular Momenta Entanglement

As mentioned earlier, the idea is to start with polarization entanglement, which can be efficiently produced with high quality, and transfer it afterwards to the transverse spatial degree of freedom. A sketch of the full setup to generate entanglement of high OAM can be seen in Fig. 3.10. In the first step, the polarization entangled bi-photon state $|\Psi\rangle^{p}$ is generated in a Sagnac-type source. Each photon is coupled separately to a SMF for Gauss-mode filtering and redirected to the two transfer setups. In each setup, the photon's polarization is transferred to the OAM degree of freedom (see Sect. 3.2), thus entanglement in the traversal mode of the photons is generated. The bi-photon state is now

$$|\Psi\rangle^{OAM} = \frac{1}{\sqrt{2}}\Big(|{+l}\rangle\,|{-l}\rangle + e^{i\varphi}\,|{-l}\rangle\,|{+l}\rangle\Big), \tag{3.9}$$

where the positions of the ket-vectors label the different photons. The next question concerns the possible upper limit of the OAM quanta. This is closely connected to the possibility of creating OAM modes as well as detecting the OAM entanglement. Latter one will be discussed in Sect. 3.5. In the following, the limitations of the presented setup in creating modes with high OAM values will be investigated.

Limitation of the Transfer Setup

The highest published value of OAM created classically with an SLM was 200 $\hbar$ [21]. A brief estimation of the limitation due to the finite resolution of the SLM

[4] The scope of the experiment in [17] is a different one. There, the focus was to increase the dimensionality of entanglement.

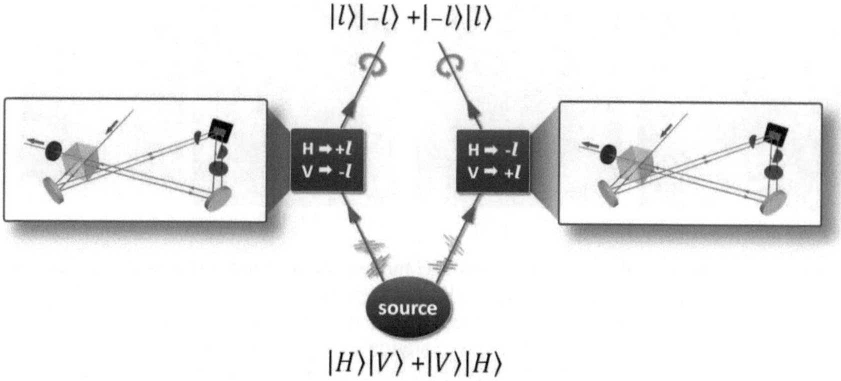

Fig. 3.10 Schematic of the setup to entangle high OAM quanta

and the expected LG mode orders can be conducted. The maximal diameter of the displayed phase mask on the SLM is 960 pixels. Assuming 5 pixels per 2π phase shift are sufficient to create a LG mode, approximately 600 such phase ramps can be drawn in the outer part of the phase mask. However, in the inner area of the phase pattern there are less pixels per 2π change available. This is the main limiting factor of the presented experiment. The emergence of the unwanted Moirè pattern due to aliasing effects reduces the production efficiency and makes even higher OAM modes impossible (see the displayed phase pattern in Fig. 3.11). However, it is important to note that this is only a technical limitation and will be overcome by advances in technology for generating higher-order modes. In the experiment a test with a laser going through the transfer setup was performed.

For values higher than $l = 10$, an additional reduction in the modal overlap was observed because the two different modes were slightly elliptical. The reason for the ellipticity was most probably the fact that the incident light at the SLM was not perpendicular to the SLM surface and the SLM surface itself is not perfectly flat. However, it is possible to correct these effects by displaying additional lenses on the SLM. To increase the circularity of the mode, a metropolis-optimization algorithm

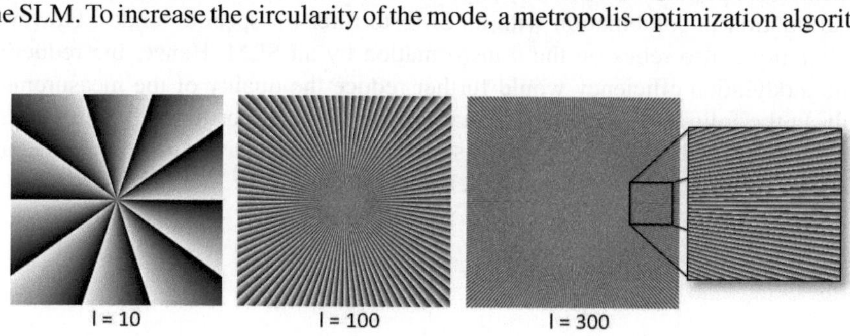

Fig. 3.11 Phase pattern for higher-order LG modes carrying high OAM quanta. Due to the finite resolution of the SLM, Moirè patterns emerge for modes with $l = 300$. However, a zoom reveals that the required azimuthal phase dependence is still being modulated by the SLM

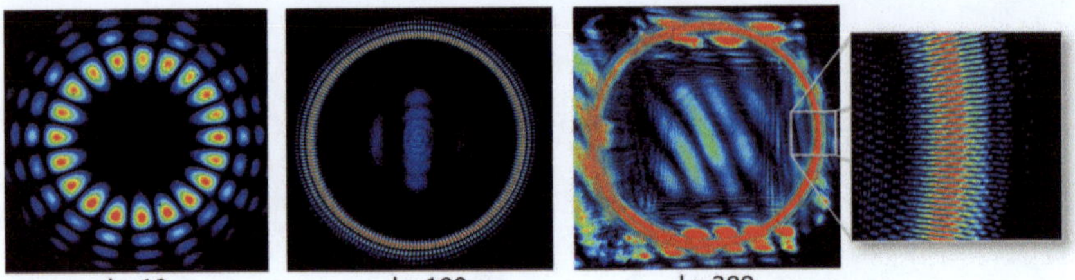

Fig. 3.12 Experimental false color images of LG mode superpositions taken with a CCD camera. Modes up to $l = \pm 100$ can be generated with very high quality. Additional structures outside of the main intensity ring are caused by imperfect modulations but carry the same amount of OAM. Intensities inside the ring already indicate a reduction in the modulation efficiency. Both effects are not relevant to later experiments because they are cut out by the measurement apparatus. For $l = \pm 300$, diffracted light from all implemented optics and mechanics, as well as from the distorted modulation interferes with the generated mode. However, the whole mode does not get destroyed. A small part of the beam (inset: zoom of structure) still shows the superposition structure and can be used to test for entanglement (at the cost of low count rates)

was implemented. The algorithm fitted an ellipse to the observed picture and reduced the ellipticity by finding the best cylindrical lenses (the free parameters were the two focal lengths of the cylindrical lenses and their rotation angle around the optical axis). For values of $l = 300$ and higher, a clear reduction of mode transformation efficiency was observed (see Fig. 3.12 for experimental results). Although the whole mode is heavily distorted by interference with diffracted light from the SLM housing or the mirror and PBS edges, a small part of the beam can still be used to test for OAM entanglement (see inset in Fig. 3.12). Hence, the presented transfer setup is limited by the SLM to 300 quanta of OAM generated per single photon.

3.5 Measurement of High Angular Momenta Entanglement

After OAM entanglement has been generated, the verification of the expected entanglement has to be performed. A very popular measurement scheme uses a holographic transformation in combination with an SMF to filter for specific modes. However, this technique also relies on the transformation by an SLM. Hence, the reduction of the modulation efficiency would further reduce the quality of the measurement result. In the following, a novel tool that probes the superposition structures of LG modes with $\pm l$ will be introduced. This method is inexpensive, fast to manufacture and powerful enough to verify entanglement.

3.5.1 Slit-Wheel for Measuring OAM States

As shown earlier, the rotation angle of a superposition of two LG modes $\pm l$ is directly linked to the phase between the modes l and $-l$. Hence, any equally weighted superposition—including two out of three MUBs—is accessible by measuring the

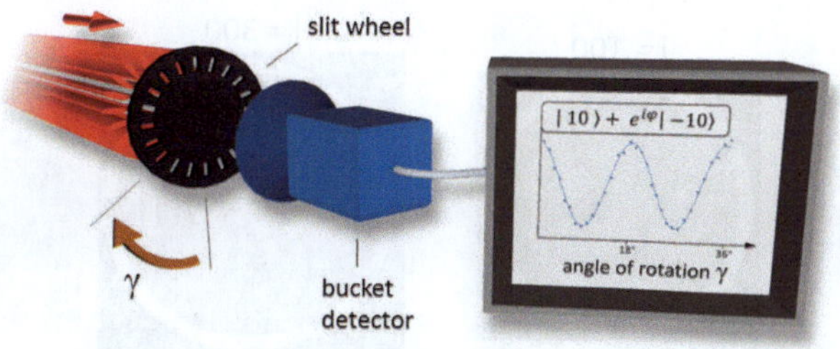

Fig. 3.13 Sketch of the measurement with a slit-wheel. The slit wheel mask is placed in a LG mode superposition. The transmitted intensity is detected by a bucket detector and varies with the rotation angle γ of the mask

intensity structure of the superposition and its angular position. Such a measurement can be performed by a so-called slit-wheel that is adapted to the lth LG mode. The slit-wheel is a mask with $2l$ transparent slits of equal spacing and width. They are arranged in a ring analogous to the superposition structure, while the rest of the mask is opaque. Light in a LG mode superposition of $\pm l$ is transmitted with a high efficiency for specific angles, namely when the slit-wheel angle γ fits to the phase φ of the superposition. If the mask is rotated by an angle $\Delta\gamma$ the transmitted intensity decreases until a minimum is obtained ($\frac{\pi}{2l}\frac{360^\circ}{2\pi}$ after the angle of the maximum transmittance). The idea can be seen in a sketch in Fig. 3.13. The theoretical formalism describing the slit-wheel measurements was developed by William N. Plick. It will be described in the Appendix A as an additional justification to the intuitive description given here. The highest precision in angular resolution, and thus in the measurement of a specific superposition, is obtained if the slits are infinitesimally narrow. Therefore, the narrower the slits, the higher the visibility in the measurement of the LG mode superposition. However, the narrower the slits, the less light that can pass through the mask, which leads to an increase in measurement time. In Fig. 3.14, a graph shows the visibility and the transmitted intensity (if the mask is placed in the maximum) depending on the ratio of distance between slits and slit width. The visi-

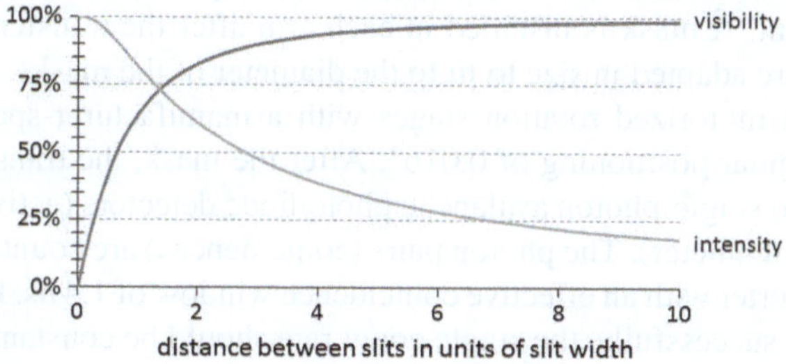

Fig. 3.14 Calculation of the maximal measurable visibility depending on the distance between slits in units of the slit width (*blue*). The narrower the slits, the bigger the ratio and the better the visibility, but the lesser the transmitted intensity (*green*, normalized to the full intensity without any masking)

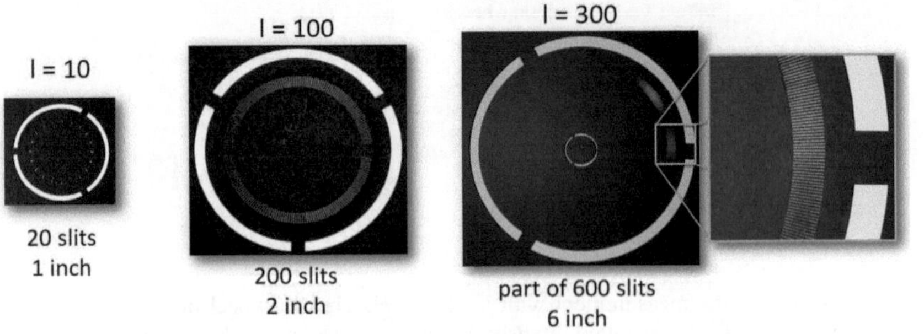

Fig. 3.15 Slit-wheels for different LG mode superpositions cut out of black paper (sizes not to scale)

bility is calculated to be the ratio between the difference and the sum of the intensity in the minimum and maximum.

For the experiment, the masks were cut out of black paper by a laser cutter. Because the laser cutter has a minimum cutting width, the masks were adapted in size for each slit-wheel (see Fig. 3.15). The masks for $l = \pm 10, \pm 100$, and ± 300 were cut out of a 1-inch, 2-inch and approximately 6-inch diameter sized piece. In the 6-inch mask only part of the whole ring was cut out because the rest of the beam was not usable. The slit width was measured to be around 300(35) μm for $LG_{\pm 10}$ and 100(10) μm for $LG_{\pm 100}$ and $LG_{\pm 300}$. The ratio of the distance between two slits and the slit width is approximately 7.1 ($l = \pm 10$), 5.7 ($l = \pm 100$), and 6.9 ($l = \pm 300$), which reduces the theoretical maximum fringe visibility to around 96.8(6), 95.0(3), and 96.6(7)%, respectively.

3.5.2 Measurements of Entanglement of High OAM Quanta

In the earlier Sect. 2.4.1, it was shown that measuring only two out of three MUBs is enough to verify non-classicality. Since angle-depend slit-wheel measurements of transmitted photons are able to accomplish that, they can be used to demonstrate the entanglement. A mask is installed in each arm after the transfer setups and the spatial modes are adapted in size to fit to the diameter of the masks. The slit-wheels are mounted in motorized rotation stages with a manufacturer-specified absolute accuracy in angular positioning of 0.016°. After the mask, the transmitted photons are focused onto single-photon avalanche photodiode detectors (active area approximately 500 μm diameter). The photon pairs (coincidences) are counted by a time-to-amplitude converter with an effective coincidence window of 1.4 ns. If the LG modes were entangled successfully, the single count rate should be constant irrespective of the angular position of both masks. In contrast to the coincidence count rate that should depend strongly on the angular positions of both masks, or more precisely, that should only depend on the difference between the angular positions (Fig. 3.16). As mentioned earlier, the underlying theory developed by William N. Plick is given in

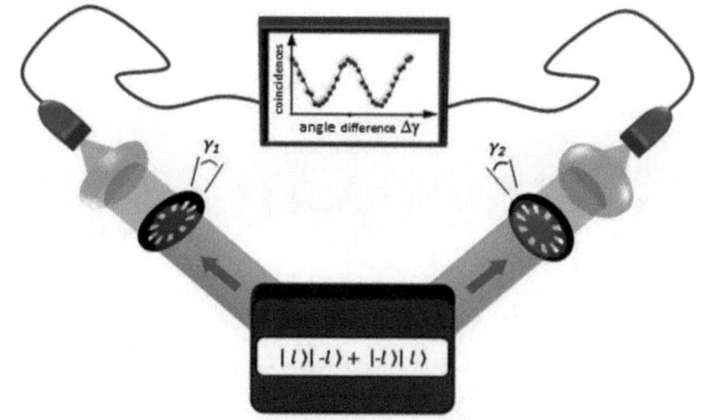

Fig. 3.16 Sketch of the measurement of entanglement with slit-wheels. After the high OAM entanglement has been generated (*blue box*), a slit-wheel is installed in each arm. Transmitted photons are detected by single-photon detectors (*green*) and coincidences are registered. The coincidence rate should only depend on the difference in angular position $\Delta\gamma = \gamma_1 - \gamma_2$ between the two masks

Appendix A, which justifies the use of the slit-wheel masks as quantum-mechanical measurements devices from a theoretical point of view.

As a first test, the flexibility of the whole setup can be demonstrated by transferring one photon to $l = \pm 10$ and the other photon to $l = \pm 100$, thus creating the state

$$|\Psi\rangle^{\pm 10/100} = \frac{1}{\sqrt{2}}\Big(|+10\rangle\,|-100\rangle + e^{i\varphi}\,|-10\rangle\,|+100\rangle\Big). \tag{3.10}$$

Apart from the general difficulty to create very high OAM values directly from the down conversion process, it would be hardly possible to obtain such an asymmetric state due to the intrinsic angular momentum conservation in the process. The resulting fringes that are characteristic for entanglement measurements can be found Fig. 3.17. Additionally, the single counts behind the first photon's mask ($l = \pm 10$) are shown. The single counts are not as flat as theoretically expected, which can be explained by misalignments of the mask with respect to the optical axis. Nevertheless, no dependence on the angular position of mask of the second photon ($l = \pm 100$) is found, thus the fringes in the coincidence rate cannot be explained by the fluctuation in the single count rate.

In a second experiment, both photons are transferred to $l = \pm 100$ to show the capability of the novel setup for creating OAM entangled photons with very high difference in their quantum number. The generated state is now

$$|\Psi\rangle^{\pm 100} = \frac{1}{\sqrt{2}}\Big(|+100\rangle\,|-100\rangle + e^{i\varphi}\,|-100\rangle\,|+100\rangle\Big). \tag{3.11}$$

Again, fringes with high visibility that only depend on the angle difference between the two masks can be observed. Already after 3.6° rotation of one mask, two full fringes corresponding to a 4π phase shift are measured. In addition, a scan over a broader angular region of one mask has been performed while the second mask is

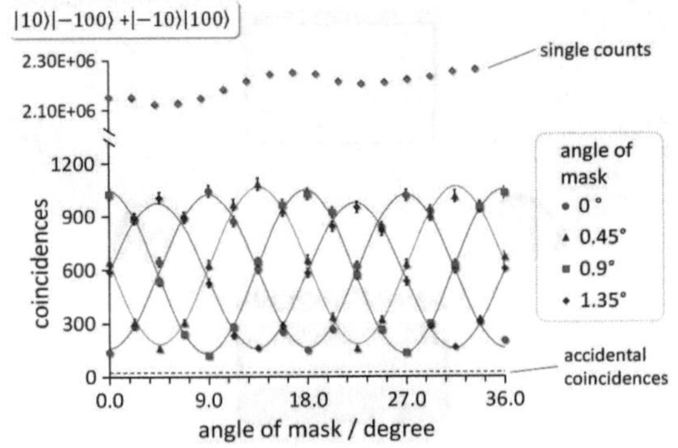

Fig. 3.17 Measured non-classical coincidence counts for entanglement between photons with $l = \pm 10$ and partner photons with $l = \pm 100$. The coincidence counts (circular points, 2 min integration time) depend on the rotation angle of the first mask (20 slits) and the second mask (200 slits). The single counts (diamond points) solely depend on the angle of the first mask. Thus, they can be ascribed to misalignment of the mask but not to explain the measured correlations. Note that the singles for all four mask angles—0°, 0.45°, 0.9° and 1.35°—are shown but they are overlaid by each other. Error bars (if big enough to be seen) are estimated from Poissonian count statistics and the line is the best fit to the theoretical $\sin^2$-function

unchanged. The expected 50 fringes for a rotation from 0° to 90° have been observed (see Fig. 3.18).

The highest value of OAM per single photon, where strong correlations were still measurable, was $l = \pm 300$ for both photons.

$$|\Psi\rangle^{\pm 300} = \frac{1}{\sqrt{2}}\Big(|{+300}\rangle\,|{-300}\rangle + e^{i\varphi}\,|{-300}\rangle\,|{+300}\rangle\Big) \tag{3.12}$$

However, the decrease in the mode transformation efficiency of the SLM (discussed in detail in Sect. 3.4.3) strongly affects the coincidence rate. Approximately only 1 coincidence count per minute was detected in the maximum. Although two correlation fringes are still visible (see Fig. 3.19), the statistical significance of the results is reduced. Moreover, as will be seen later, the introduced witnesses can no longer be evaluated and a more subtle way of demonstrating the non-separable behavior has to be discussed.[5]

The visibility in both accessible MUBs can be calculated from the minima and maxima of the recorded fringes. Thus the introduced entanglement witness $\hat{W}_2$ from Sect. 2.4.1, which requires two visibilities each derived from two anti-correlated fringes, can only be evaluated for the generated states $|\Psi\rangle^{\pm 10/100}$ and $|\Psi\rangle^{\pm 100}$.

[5] In an earlier investigation [1] an erroneous witness was introduced and used to show entanglement. Here, we show that only for the highest OAM quanta, namely the state $|\Psi\rangle^{\pm 300/300}$, a direct evaluation of an entanglement witness is not tenable anymore. However, it is still possible to demonstrate that the measurements cannot be described by a separable state, hence the claims from [1] remain the same.

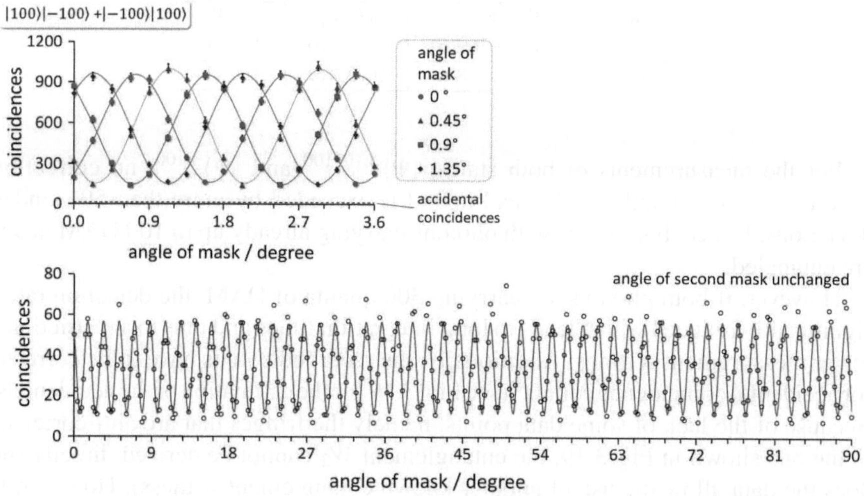

Fig. 3.18 Measured coincidence counts when both photons are transferred to $l = \pm 100$. The *upper graphs* shows the typical entanglement feature that the coincidence counts (9 min integration time) only depend on the angle difference between the two masks with 200 slits (error bars from Poissonian count statistics). The *lower graphs* shows the coincidence counts for a broad scan of one mask while the other mask is kept at a fixed angle (error bars are not shown for the sake of clarity). After 90° of rotation 50 fringes are observed. Each data point corresponds to 30 s of measurement time. All lines are the best fit to $\sin^2$-functions

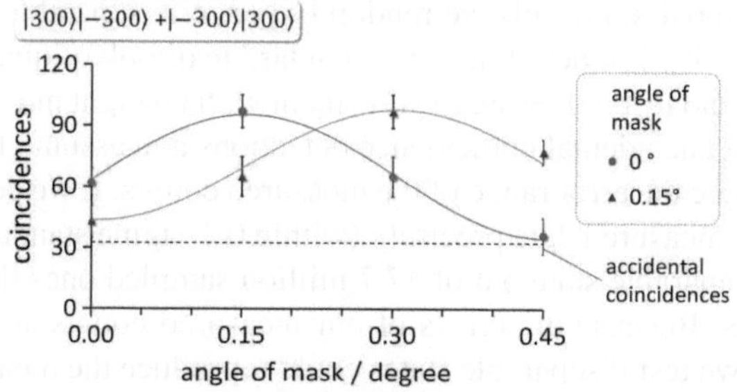

Fig. 3.19 Measured entanglement fringes when both photons are transferred to $l = \pm 300$. Here, the count rate decreased significantly to around 1 coincidence count per minute. Error bars assume Poissonian count statistics and the line shows the best $\sin^2$-fit

For W_2 the classical bound was found to be

$$W_2 = vis_{\gamma_{1/3}} + vis_{\gamma_{2/4}} \quad \begin{cases} \leq 1 & \text{separable} \\ > 1 & \text{entangled} \end{cases} \tag{3.13}$$

Here, the indices $\gamma_{1/3,2/4}$ label the angular positions of the mask, which is kept at a fixed angle while one fringe is measured. In the following table, all measured visibilities as well as the calculated witness values are shown for these measurements.

state	$vis_{\gamma_{1/3}}$	$vis_{\gamma_{2/4}}$	W_2
$\|\Psi\rangle^{\pm 10/100}$	0.769(6)	0.725(6)	1.49(1)
$\|\Psi\rangle^{\pm 100}$	0.728(6)	0.732(6)	1.46(1)

For the measurements of both states, $|\Psi\rangle^{\pm 10/100}$ and $|\Psi\rangle^{\pm 100}$, no corrections have to be applied and the classical bound is exceeded by more than 50 standard deviations. Hence, both states with photons carrying already up to 100 OAM quanta are entangled.

However, if both photons are carrying 300 quanta of OAM, the detection rate is significantly reduced which leads to long integration times and thus to a restriction to fewer measurement settings. Additionally, the recorded counts have to be corrected for accidental coincidence counts[6] (see Fig. 3.19 for the estimation of the accidentals). Because of the lack of some data points, namely the fringes that are anti-correlated to the one shown in Fig. 3.19, the entanglement W_2 cannot be derived directly (nor does the data allow the use of another known entanglement witness). However, by assuming the same behavior for the missing two anti-correlated fringes the witness value would be 1.6(3), which would exceed the classical bound as well. Although this is only an indication, the assumption might be justified since the correlation fringes were recorded from a random starting point on and show the typical behavior of entangled systems, where the coincidence counts only depend on the relative angle between the two masks.

To further strengthen this indication, we compare our results to separable states describing photon pairs. For this we randomly generate separable states (including mixed states, described in detail in every standard textbook of quantum mechanics, e.g. [22] or [23]) and test if they are in agreement with all eight measured data points. Again, we subtract accidental coincidence detections and assume Poissionian count statistics to estimate the error range of the measured counts. If we constrain the states to agree with our measured data precisely (within 0.1-sigma standard deviation), we do not find any separable state out of 47.7 million sampled ones that can reproduce all measurements. Because the errors of our measured counts are large due to the low count rates, we test if separable states could reproduce the measured data within a 1-sigma and 2-sigma standard deviation error range. For both cases, no separable state was found to be in agreement with all measurements and still reproduce the sum of the two measured visibilities. This further demonstrates that it is highly unlikely that a separable state could explain our measurements. However, 362 and 121139 separable states were found to agree within a 1-sigma and 2-sigma standard deviation interval, respectively (out of 52.7 and 19.1 million sampled ones). To demonstrate that graphically, we plotted the distribution of the sum of the two visibilities for all randomly generated states that could explain all measured counts within the described error intervals (see Fig. 3.20). Again, we find that our data strongly indicates the successful generation of entanglement with 300 quanta of OAM.

[6]The accidental coincidences are calculated according to $accs = S_1 * S_2 * \tau$, where S_i labels the singles behind both slit-wheels $i = 1, 2$ and τ stands for the coincidence window.

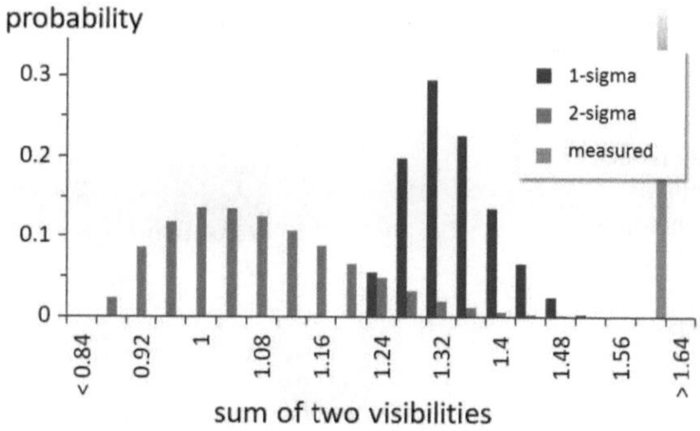

Fig. 3.20 Probability distribution of the sum over two visibilities for separable states that are in agreement with the correlations measured for the state $|\Psi\rangle^{\pm 300}$. <0.001 % of the separable states agree with the measurements within a 1-sigma interval (*blue columns*). If the separable states are only constrained to lie within a 2-sigma interval, around 0.6 % are in agreement with the measurements and the distribution is broadened (*red columns*). None of them reproduces the actual measured visibilities (*green column*), hence both distributions strongly indicate that the data arises from entangled photons

In addition to the analysis discussed above, a test was performed where only one photon was transferred to $l = \pm 300$ and the other was kept polarization-encoded (hybrid entanglement, see Sect. 3.5.3 for more information). Here, the count rate was high enough such that all four correlation extrema were measured and the witness W_2 can be evaluated:

state	$vis_{D/A}$	$vis_{R/L}$	W_2
$\|H\rangle\|-300\rangle + \|V\rangle\|+300\rangle$	0.8138(4)	0.7765(4)	1.5903(5)

The indices *D*/*A*, *R*/*L* stand for the polarization measured for the unchanged photon. The data exceeds the classical bound by more than 1000 standard deviations. This demonstrates that the single photons can carry 300 quanta of OAM and still be entangled. Moreover, it shows that the strong indication of entanglement between two photons each carrying $300\,\hbar$ is well-founded and the missing violation of the described entanglement witness only appears due to the lower count rates and not due to any fundamental limits.

3.5.3 Enhancement of Angular Sensitivity

Apart from the fundamental interest of entanglement of very high quantum numbers, it is also possible to demonstrate the use of very high OAM for remote sensing applications. Similar to the very last experimental test in the previous section, only one photon is transferred to LG modes while the other one kept in the polarization

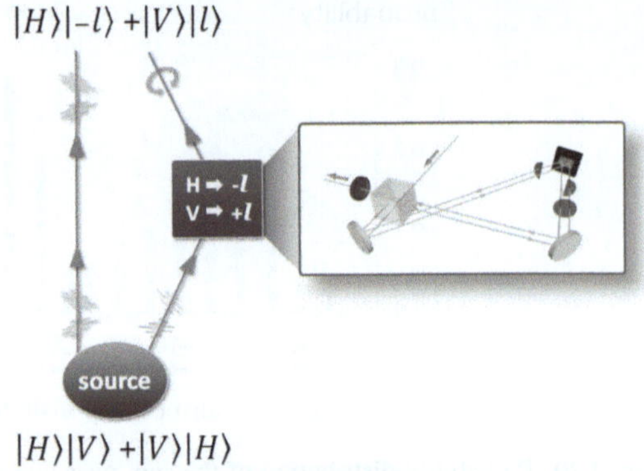

Fig. 3.21 Sketch of the setup to create hybrid-entangled states between polarization and transverse spatial LG modes. Compared to the setup of Fig. 3.10, only one photon is transferred (*right*) while the other is kept in its polarization state (*left*)

state (see Fig. 3.21). The photon pair is then in a so-called hybrid-entangled state $|\Psi\rangle^{hybrid}$ between polarization and the orbital angular momentum

$$|\Psi\rangle^{hybrid} = \frac{1}{\sqrt{2}}\Big(|H\rangle\,|-l\rangle + e^{i\varphi}\,|V\rangle\,|+l\rangle\Big). \tag{3.14}$$

hybrid-entangled states are very interesting because they can be used to remotely measure an angular rotation with a precision that is increased by a factor l relative to the situation when only polarization-entangled photon pairs are used (a similar idea of taking advantage of high OAM values was described earlier in [24]). As already mentioned above, the physical orientation γ of the mask which the transferred photon passes, and the phase φ of the superposition state that the mask selects, are connected via the equation $\gamma = \frac{\varphi}{2l}\frac{360°}{2\pi}$. In the experiment this leads to an enhanced change in the coincidence count rate for high OAM values if the slit-wheel is rotated (see Fig. 3.22). The fundamental limit in the precision of remote angular position measurements is

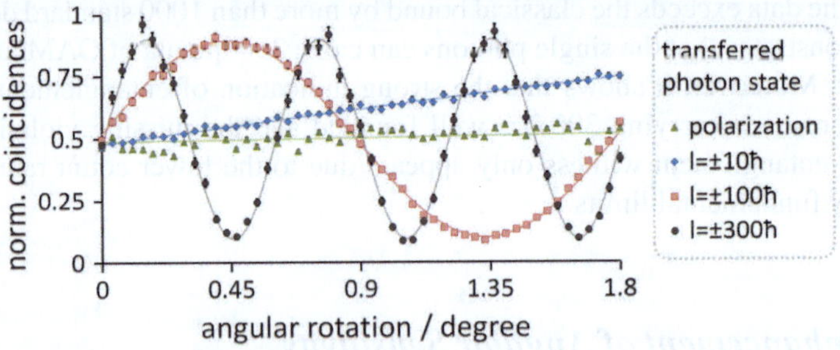

Fig. 3.22 Normalized coincidence count rates for different hybrid-entangled states. One photon is kept in its polarization state and projected into diagonal polarization. The second photon is either kept polarization-encoded while the polarizer is rotated (*green triangles*), or it is transferred to $l = \pm 10$ (*blue diamonds*), $l = \pm 100$ (*red squares*), or $l = \pm 300$ (*black circles*) while the appropriate mask is rotated. The errors are estimated assuming Poissonian count statistics

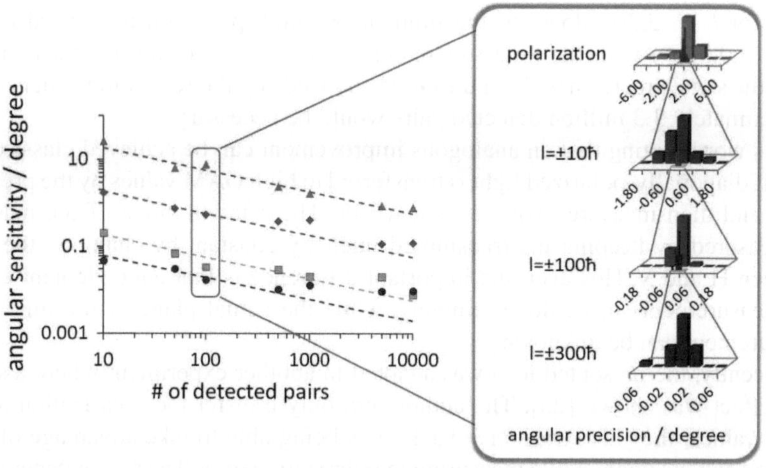

Fig. 3.23 Angular sensitivity depending on the number of photons (at the steepest slope of the fringes seen in Fig. 3.22). *Left* Depending on the number of photons involved in the measurement of an angular position, the statistical fluctuations limit the precision of the measurement. The *dashed lines* represent the theoretically expected sensitivities, assuming 100 % visibility and Poissonian fluctuations. The data points depict the measured fluctuation including all experimental instabilities. *Right* For each setting 20, different angles of randomly rotated masks were measured remotely by the presented scheme (main text). By transferring one photon to $l = \pm 300$ it is possible to measure the angular position down to approximately 0.02°

given in the experiment by the fluctuation of the coincidence count rate. Hence, the statistical fluctuation if only a low number of photons has to be used sets a rather high upper limit on the precision of angular measurements (see Fig. 3.23). Because a high OAM quantum number l acts as an inverse scaling factor, the remotely measured angular position of an object can be enhanced. In other words, a small rotation of the mask of the transferred photon can be detected easily. The change can be undone by the rotation of a polarizer placed in the path of the unchanged photon, from which the actual angular rotation of the mask can be evaluated. Due to the inverse scaling, the angles that have to be measured with the polarizer are l times bigger than the angle of the mask. Thus, hybrid entanglement acts as a gear shift between the two rotations, the one from the polarizer and the rotation from the slit-wheel.

To test the described enhancement in a real experiment, the most sensitive point to angular instabilities, namely the steepest slope of the coincidence fringe, is chosen. It is attempted to keep the count rate constant at only 100 detected pairs per second to show the capability at a low number photons. For different OAM values, the corresponding mask (or polarizer if the second photon is not changed) is randomly rotated. After every rotation, the induced change in the coincidence rate is undone by a rotation of the polarizer in the path of the polarized first photon. In Fig. 3.23 on the right, histograms are shown consisting of 20 angular position measurements for each arrangement. It was possible to determine the angles of the randomly rotated mask for $l = \pm 300$ down to the limit of our high precision motorized rotation stage ($\pm 0.016°$) with the polarizer mounted in a low-precision mount ($\pm 1°$). To reach the same precision without the OAM-induced angular resolution enhancement, approximately 3.3 million detected pairs would be necessary.

It is worth noting that an analogous improvement can be achieved classically as well, if diagonally-polarized light is transferred to high OAM values by the presented setup and then measured with the slit-wheel. The rotation angle of the mask can be measured by keeping the transmitted intensity constant by changing the phase between H and V. However, the important difference is that entanglement enables the measurements to be done remotely, while the actual plane or the time of the measurement can be unknown.

Recently, the presented idea was adapted in another experiment where a similar gear-effect was shown [25]. The authors not only transfer the polarization to high OAM values, they also transfer it back, thus being able to take advantage of the l-induced enhancement while measuring in polarization only. The recent paper as well as the presented results demonstrate that high quanta of OAM can be used not only for fundamentally interesting experimental tests but in angular sensing applications.

However, the measurement principle with the slit-wheel only works for LG mode entanglement, while the presented setup is flexible in creating entanglement of any transverse spatial structure. Hence, another way to detect the entangled photon pairs is required. One novel possibility will be discussed in the next chapter: coincidence-imaging with a triggered, intensified CCD camera.

References

1. Fickler, R., et al. (2012). Quantum entanglement of high angular momenta. *Science*, *338*, 640.
2. Jesacher, A. (2009). *Applications of spatial light modulators for optical trapping and image processing*. Ph.D. thesis, Leopold-Franzens University.
3. Fedrizzi, A. (2008). *Fundamental experiments with a high brightness source of entangled photons*. Ph.D. thesis, University of Vienna.
4. Kim, T., Fiorentino, M., & Wong, F. N. C. (2006). Phase-stable source of polarization-entangled photons using a polarization Sagnac interferometer. *Physical Review A*, *73*, 012316.
5. Fedrizzi, A., Herbst, T., Poppe, A., Jennewein, T., & Zeilinger, A. (2007). A wavelength-tunable fiber-coupled source of narrowband entangled photons. *Optics Express*, *15*, 15377.
6. Żukowski, M., & Pykacz, J. (1988). Bell's theorem: Proposition of realizable experiment using linear momenta. *Physics Letters A*, *127*, 1.
7. Ramelow, S., Ratschbacher, L., Fedrizzi, A., Langford, N. K., & Zeilinger, A. (2009). Discrete tunable color entanglement. *Physical Review Letters*, *103*, 253601.
8. Nagali, E., et al. (2009). Quantum information transfer from spin to orbital angular momentum of photons. *Physical Review Letters*, *103*, 013601.
9. Galvez, E.J., Nomoto, S., Schubert, W. & Novenstern, M. (2011). Polarization-spatial-mode entanglement of photon pairs. In *International Conference on Quantum Information*, Washington, Optical Society of America.
10. Leggett, A. J. (2002). Testing the limits of quantum mechanics: Motivation, state of play, prospects. *Journal of Physics: Condensed Matter*, *14*, R415.
11. Aspelmeyer, M., Meystre, P., & Schwab, K. (2012). Quantum optomechanics. *Physics Today*, *65*, 29.

12. Mair, A., Vaziri, A., Weihs, G., & Zeilinger, A. (2001). Entanglement of the orbital angular momentum states of photons. *Nature, 412*, 313.
13. Miatto, F. M., Yao, A. M., & Barnett, S. M. (2011). Full characterization of the quantum spiral bandwidth of entangled biphotons. *Physical Review A, 83*, 033816.
14. Jack, B., et al. (2009). Precise quantum tomography of photon pairs with entangled orbital angular momentum. *New Journal of Physics, 11*, 103024.
15. Pires, H. D. L., Florijn, H. C. B., & van Exter, M. P. (2010). Measurement of the spiral spectrum of entangled two-photon states. *Physical Review Letters, 104*, 020505.
16. Romero, J., Giovannini, D., Franke-Arnold, S., Barnett, S. M., & Padgett, M. J. (2012). Increasing the dimension in high-dimensional two-photon orbital angular momentum entanglement. *Physical Review A, 86*, 012334.
17. Krenn, M., et al. (2014). Generation and confirmation of a (100 × 100)-dimensional entangled quantum system. *Proceedings of the National Academy of Sciences, 11*, 6243.
18. Svozilík, J., Peřina, J, Jr, & Torres, J. P. (2012). High spatial entanglement via chirped quasi-phase-matched optical parametric down-conversion. *Physical Review A, 86*, 052318.
19. Campbell, G., Hage, B., Buchler, B., & Lam, P. K. (2012). Generation of high-order optical vortices using directly machined spiral phase mirrors. *Applied Optics, 51*, 873.
20. Shen, Y., et al. (2013). Generation and interferometric analysis of high charge optical vortices. *Journal of Optics, 15*, 044005.
21. Jesacher, A., Fürhapter, S., Maurer, C., Bernet, S., & Ritsch-Marte, M. (2006). Holographic optical tweezers for object manipulations at an air-liquid surface. *Optics Express, 14*, 6342.
22. Nielsen, M. A., & Chuang, I. L. (2010). *Quantum computation and quantum information*. Cambridge: Cambridge University Press.
23. Audretsch, J. (2008). *Entangled systems*. New York: Wiley.
24. Jha, A. K., Agarwal, G. S., & Boyd, R. W. (2011). Supersensitive measurement of angular displacements using entangled photons. *Physical Review A, 83*, 053829.
25. D'Ambrosio, V., et al. (2013). Photonic polarization gears for ultra-sensitive angular measurements. *Nature Communications, 4*, 2432.

Chapter 4
Coincidence Imaging of Spatial Mode Entanglement

4.1 Preamble

The following chapter is based on the publication "Real-time imaging of quantum entanglement" [1].

As outlined in the previous section, the transfer setup is very flexible in creating two-dimensional transverse spatial mode entanglement. The investigation of such modes and their complex spatial patterns requires a detection scheme with a high spatial resolution. Masking the modal structure (as shown in the previous section) or scanning a single pixel detector enables the detection of spatial structures. However, these schemes come at a certain cost. Masking the beam and thereby measuring its transverse spatial structure needs to be adapted to the specific mode properties that are under investigation. Hence, it is not flexible for use with a larger variety of modes without adapting the shape of the mask (if possible at all). On the contrary, scanning a single pixel detector does not need to be adapted to the measured modal structure. However, the higher the resolution, the less efficient the detection is, which leads to a drastic increase in time consumption.

The rapid progress in imaging technologies over recent years has made charged coupled device (CCD) cameras an interesting alternative for the detection of single photons in quantum optics experiments. The available CCD chips have a huge number of pixels, each sensitive to single-photon detection events (either by pre- or post-intensifying the signal). Therefore, the spatial information for investigating the structure is directly accessible with a very high resolution.

Electron multiplied CCD cameras have attracted attention recently because of their high quantum efficiencies of more than 90 %.[1] They have been utilized for

[1] The quantum efficiencies (QE) given for CCD cameras should not be confused with the detection efficiencies of single-photon detectors, which can be as high as 60–70 % for avalanche photodiodes (APD). The QE for CCD chips is the ratio between incident photons and electrons generated due to the photon impact. It is often called the spectral response because of its wavelength dependence. Compared to APDs, the detection efficiency is lower because not all electrons might be detected. Other effects, such as fill factor of the chip or absorption of the photon at the gate electrodes, have to

R. Fickler, *Quantum Entanglement of Complex Structures of Photons*,
Springer Theses, DOI 10.1007/978-3-319-22231-8_4

demonstrating non-classical correlations of photons produced via spontaneous parametric down conversion (SPDC) [2–5]. However, such cameras have a restrictive downside because they only allow relatively long exposure times in the range of a few μs. This makes it necessary to sum over many sparse images (images with only two correlated photons at best) to detect correlations. Furthermore, it is unfeasible to use them for coincidence-imaging of entanglement where precise fast timing within a few nanoseconds is essential. In contrast, a different type of single-photon sensitive CCD camera, namely an intensified CCD camera (ICCD), is not limited to μs-timing. This suggests a use in quantum optical experiments.

4.1.1 Intensified CCD Cameras

As mentioned above, there are different ways to increase the sensitivity of CCD chips to the single-photon level. For intensified CCD (ICCD) cameras, the signal of a single photon is intensified before it is imaged with the CCD chip.[2] The amplification of the signal of a single photon is done by the so-called intensifier (see Fig. 4.1). This device principally consists of three important elements: a photocathode to convert photons into electrons, a micro-channel-plate (MCP) to multiply the number of electrons and a phosphor plate to convert the electron avalanche to photons. Hence the whole process can be described by the following chain of reactions. First an incoming photon kicks out a photo-electron at the photocathode. Then, the photo-electron is multiplied in an avalanche process in the MCP by applying a high voltage. The MCP can be regarded as many photomultiplier tubes mounted in parallel in a honeycomb structure next to each other. Afterwards, the generated electrons are accelerated towards the phosphor plate, where they produce a flash of many photons. This photonic signal is then imaged via a fiber optical coupling by the CCD chip. The biggest advantage of this procedure is that the CCD chip does not detect any signal without any voltage at the MCP. Thus, a low noise level can be achieved without the need for deep cooling of the chip. Moreover, the MCP can be gated very fast ($\sim$2 ns) and with a very high precision ($\sim$10 ps), which additionally increases the signal-to-noise ratio and makes it possible to use such ICCD cameras in a coincidence-imaging scheme.[3] The whole intensifying process comes at a cost of detection efficiency due to additional losses and non-perfect transformation efficiencies. Nevertheless, ICCD cameras have been used to image non-classical effects for photons produced in an SPDC process [6–9].

The ICCD camera used for the experiments described in this chapter is a test-camera from the distributor *LOT-QuantumDesign* (Andor iStar A-DH334T-18F-03) and has spatial resolution of 1024 × 1024 pixels (effective pixel size 13 × 13 μm).

(Footnote 1 continued)
be considered as well. For more information see the web page of one of the biggest manufacturers of scientific CCD cameras, Andor (www.andor.com).

[2]In contrast, EMCCD cameras amplify the signal of the CCD chip afterwards by an electron multiplying process.

[3]All specifications of the ICCD camera have been taken from the manufacturer.

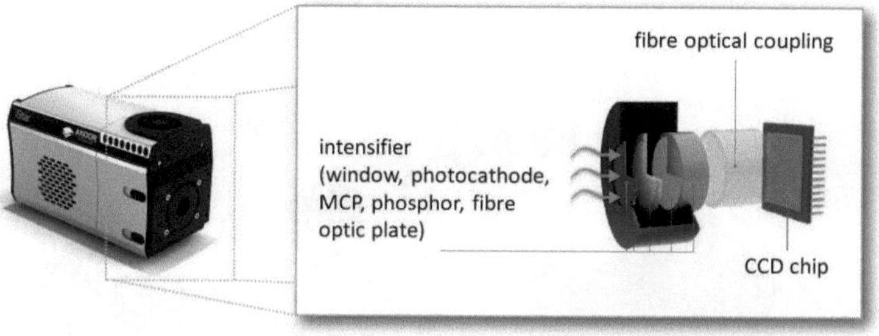

Fig. 4.1 Photo of the ICCD camera used in the presented experiments. *Right* Schematic showing the construction of the intensifier. The functional parts of the intensifier are the photocathode (*red*), the micro channel plate (MCP, *grey*) and the phosphor screen (*green*). Both images are taken from the Andor webpage (www.andor.com)

However, the actual spatial resolution is reduced by the fact that the MCP channels are bigger than the CCD pixels, which leads to single-photon events that are distributed over a few pixels (up to approx. 9 pixels). The quantum efficiency of the camera is only around 3 % at 810 nm wavelength. Other types (see following Chap. 5) offer efficiencies of up to 20 % in the wavelength regime of down conversion photons, which shortens the measurement time and increases the signal-to-noise ratio. The minimal gating window (which can be compared to the coincidence window mentioned in the previous chapter) of the intensifier was limited to 5 ns with an additional reduction of the efficiency by 60 % due to the rising time of the intensifier. The maximum repetition rate was 500 kHz. If the maximum repetition rate is used, the gated events cannot be imaged in separated frames due to limitations of the read out frame rate (maximal 4 frames per second). However, it is possible to accumulate many gated detections of single photons on the CCD chip. Then, the registered images contain many single-photon events added up to show a specific intensity structure. In the following section, a method to use the intensity images was developed, while still being able to measure the approximate photon number per spatial region. Thus, the ICCD camera could be used to image effects of quantum entanglement.

4.2 Triggered Coincidence-Imaging of Hybrid Entanglement

To test the possibility of measuring spatial structures of single-photon events with the ICCD camera, a hybrid-entangled state $|\Psi\rangle^{hyb}$ is generated (as explained in the previous Sect. 3.5.3)

$$|\Psi^{hyb}\rangle = a\,|H\rangle\,|spM_1\rangle + e^{i\varphi}\cdot b\,|V\rangle\,|spM_2\rangle\,, \tag{4.1}$$

where a, b, and φ are real, $a^2 + b^2 = 1$, and spM_j labels the generated spatial modes $j = 1, 2$ of the transferred photon. To attain the timing information when

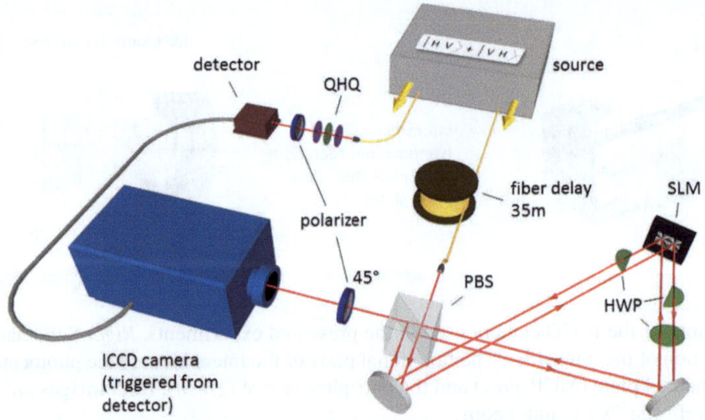

Fig. 4.2 Sketch of the setup for triggered coincidence-imaging of hybrid entanglement

the ICCD has to be gated, the detection of the first, unchanged photon is used as a trigger. Once the signal is generated by a single-photon detector and brought to the ICCD camera, there is an adjustable insertion delay[4] of a minimum of 35 ns until the intensifier is gated and able to image the transferred photons. Not only the insertion delay has to be taken into account but the additional delay of the electronics of the single-photon detector and the time-of-travel of the trigger signal until it arrives at the ICCD. Therefore, a fiber of 35 m length is introduced in the arm of the second photon. Delaying the second photon even more than necessary (around 175 ns if the speed of light is assumed to be $2 \cdot 10^8$ km/s in fiber) allows to account for mentioned delays and for the use of the adjustable, electronic delay of the camera. This can be set with a precision of 10 ps, allowing the gating window to be adapted perfectly to the arriving time of the photon. For each trigger photon detection, the ICCD camera is gated for 5 ns, thereby only registering the partner photon of the trigger, hence the name "coincidence-imaging". To visualize the effect of entanglement, a polarization measurement of the trigger photon is included. A sketch of the whole setup is shown in Fig. 4.2.

4.2.1 Transverse Spatial Photon Counting in Coincidence Images

As a first test, the exposure time was adjusted to only 100 ms for each image, which led to approximately 20 single-photon detections per frame. By investigating only

[4]The ICCD camera has a minimum insertion delay that is specified to be 19 ns. However, this is not adjustable via the electronics of the camera. Therefore, the slightly bigger but adjustable delay of 35 ns was chosen to find the best coincidence window (explained later). Moreover, an additional fiber easily delays the second photon by that time span.

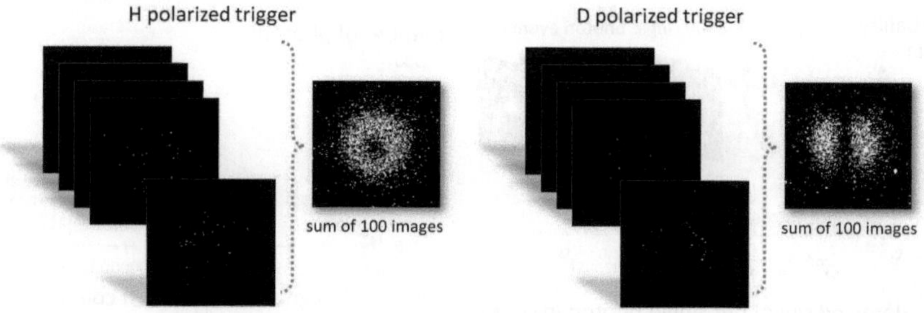

Fig. 4.3 Triggered coincidence images with very short exposure time (100 ms). In each frame only around 20 single-photon events and thus no modal structure can be found. After adding up 100 frames, the expected intensity structure of the remotely created LG mode and its superposition can be recovered

one image, no structure can be found because there are not enough detected events to reveal the transverse spatial distribution. If a sum over many images is performed, the LG mode structure and its superposition can be recovered. A few example images, as well as the sum over 100 images, can be seen in Fig. 4.3 for horizontal and vertical trigger polarization. From these images, the detected photon number per angular region could be evaluated and used to test for entanglement with an entanglement witness (analogous to the idea in the previous Chap. 3). However, to perform more complex measurements, which require more settings, this procedure would be very time consuming, mainly due to the low read out rate of 4 frames per second. Extending the exposure time to increase the number of registered photon events per image and thereby reducing the measurement time leads to an intensity pattern built by many single photons. In this intensity structure, the single-photon events mostly overlap, which makes direct counting impossible. Therefore, a method was developed to recover the approximate photon number and its error margin per transverse spatial region from a single intensity image.

4.2.2 Evaluating Photon Numbers from Coincidence Intensity Images

The registered intensity at the ICCD camera is read out as a signal in "counts per pixel". Since this response of the ICCD camera depends linearly on the photon number, the number of photons can be directly evaluated. To measure this response, an analysis of every detected photon from the sparse images shown above was performed. The responses of around 5800 photon events were considered in order to get a statistically significant value.

The procedure for the analysis performed on each image is described in the following. First, the camera-induced readout noise was subtracted from the signal registered in each pixel. Second, signals induced by real photon events were identified. A signal

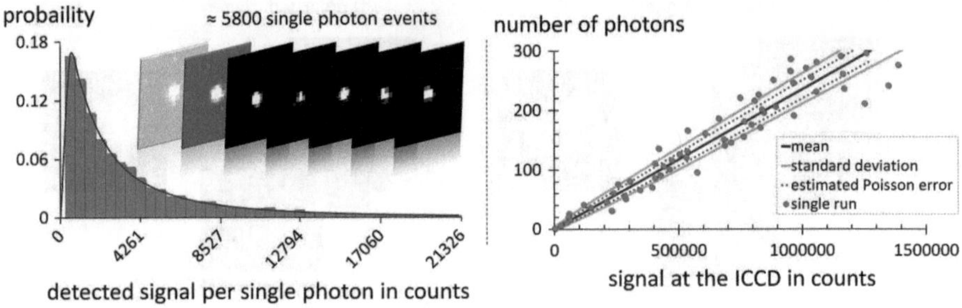

Fig. 4.4 Characterization of the ICCD camera to evaluate the photon number from registered intensities. *Left* From approximately 5800 single-photon detection (small insets) the ICCD signal (registered counts) per single photon was determined and found to be around 4200 counts per single-photon event. The histogram shows the probability distribution. The fit was conducted using a log-normal function. *Right* Using the fitted function, a Monte-Carlo simulation of count values for every reasonable photon number was performed 50,000 times. The simulated mean value (*red line*) agrees with the measured one (note that it was constructed to be like that). Because the simulated standard deviation (*orange line*) was bigger than the usual Poissonian estimation (*dashed line*), it was used in all following entanglement tests

was counted as a photon if a pixel value of more than 5 standard deviations above the background fluctuations was found. In a third step, the sum over all signal counts of the contiguous pixel array around that photon was taken as the response of the ICCD camera to the impact of one photon. The sum of all signal counts from the pixel array was utilized, since the photons are spread over a few pixels because of the different resolutions of the intensifier and the CCD chip (the spread can be seen in the insets in Fig. 4.4). From all detected photon impacts the mean value was taken as the conversion factor between the actual registered signal at the ICCD and the corresponding photon number. Hence, the measured intensity within a certain region of the image can be related to a photon number detected in that region. However, this procedure might introduce additional errors such that Poissonian count statistics do not hold to estimate the error margin.

A Monte Carlo simulation to evaluate the error margin for each photon number was performed based on the probability distribution obtained from the measurements. A very good fit to the resulting histogram of the distribution (Fig. 4.4 left side) was found to be a log-normal probability function. Using the fitted log-normal distribution, 50,000 possible signal counts were simulated for photon numbers up to 300. The simulated average signal of each photon number resembles the one obtained from the single-photon measurements. The standard deviation of the simulation is bigger than the one estimated from Poissonian statistics. Therefore, in all following measurements, the simulated error margin of each photon number was used. In Fig. 4.4 on the right side, the simulated correspondence between registered signal counts at the ICCD and the actual photon number can be found.

The method introduced above enables fast coincidence-imaging of intensity structures built by single-photon events, that depend on the measurement of the trigger photon. Once the particular ICCD model is characterized, it is possible to spatially analyze any recorded intensity image without the need for individual counting of single photons. The time consuming data acquisition of many sparse images can be avoided. Hence, the coincidence intensity images can be used directly to test for quantum effects like entanglement.

4.3 Witnessing Entanglement in Coincidence Images of LG Modes

As a first test of the coincidence-imaging technique introduced above, hybrid entanglement between polarization and the first-order LG mode was generated and analyzed. The created state reads as follows

$$|\Psi^{hyb}\rangle_{LG_{\pm1}} = a\,|H\rangle\,|LG_{+1}\rangle + e^{i\varphi}\cdot b\,|V\rangle\,|LG_{-1}\rangle\,, \tag{4.2}$$

where a, b, and φ are real ($a^2+b^2=1$) and $LG_{\pm1}$ labels the Laguerre-Gauss mode to which the second photon was transferred.[5] A pump power of 10 mW was used to generate approximately 400,000 fiber-coupled and detected polarization-entangled pairs per second in all the following measurements. An integration time of 3 s and a gating window of 5 ns led to enough coincidence detections of the transferred photons. If the camera was set to a wrong delay, only a few residual events were registered (noise). A signal-to-noise ratio of around 75:1 (see Fig. 4.5) was found. The residual—accidental—events appear mainly because of the high triggering rate (~400 kHz) and the resulting thermal noise of the intensifier or afterglow of the fluorescing phosphorous screen. These might be suppressed further by using a shorter gating time at the cost of lower efficiency.

4.3.1 Triggered Coincidence-Imaging of the Bloch Sphere for LG Modes

If a polarizer is placed in the arm of the trigger photon, thereby filtering a specific polarization, the entangled partner photon appears in the corresponding LG mode structure (see Fig. 4.6). To visualize the effect of entanglement, the polarization measurement of the trigger photon can now be varied while the spatial structure of its partner photon is being imaged. A scan through all possible polarization states of the trigger photon (the whole Poincaré sphere) results in remote coincidence-imaging of the LG-mode Bloch-sphere of the transferred photon. Because the measurement time of each spatial mode is only 3 s, the influence of the polarization measurement

[5] Instead of labeling the LG modes by their OAM content $\pm l$, as was done in the last chapter, the abbreviation $LG_{\pm l}$ will be used in this section. This is done to emphasize that the focus lies on the measurement of the transverse spatial mode rather than on its OAM content.

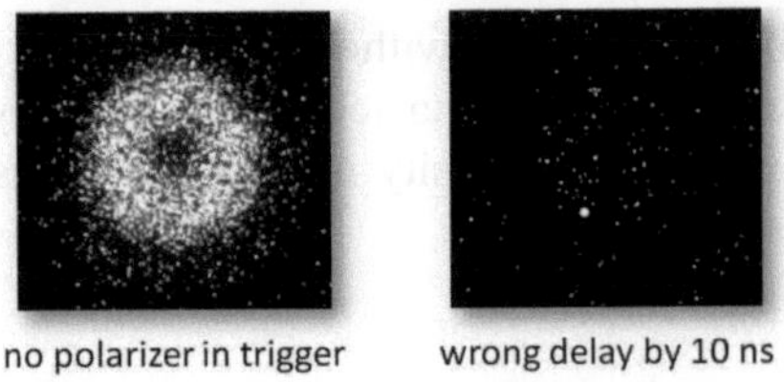

Fig. 4.5 *Left* No polarizer was put in the arm of the trigger photon. Thus, a statistical mixture of all $LG_{\pm 1}$ modes is imaged. *Right* The delay was changed by 10 ns to image the residual (accidental) events. The signal-to-noise ratio was found to be on the order of 75:1

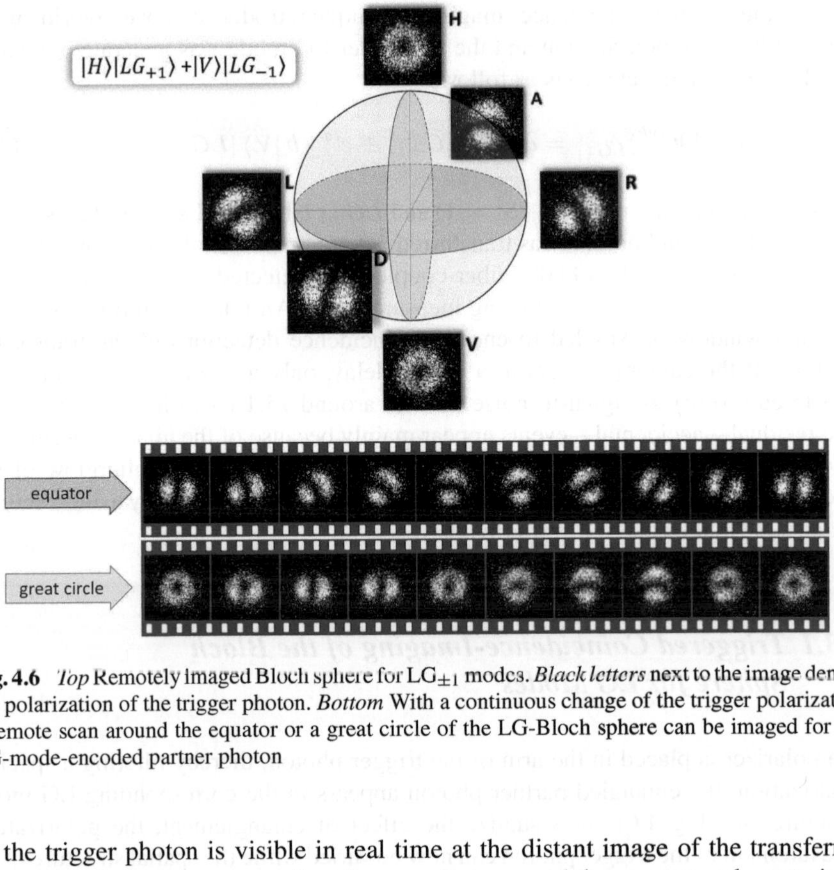

Fig. 4.6 *Top* Remotely imaged Bloch sphere for $LG_{\pm 1}$ modes. *Black letters* next to the image denote the polarization of the trigger photon. *Bottom* With a continuous change of the trigger polarization a remote scan around the equator or a great circle of the LG-Bloch sphere can be imaged for the LG-mode-encoded partner photon

of the trigger photon is visible in real time at the distant image of the transferred photon. Both the sphere as well as the snapshots around the equator and great circle can be found in Fig. 4.6. The results explained in the last section can also be seen in a youtube-video (http://www.youtube.com/watch?v=wGkx1MUw2TU).

Entanglement is already visible in this series of images. The high-contrast minima and maxima do not vanish if the polarization measurement of the trigger is changed. The extrema shift with the change of the polarization angle measured on the trigger photon. Although this visual observation already intuitively confirms the successful generation of hybrid entanglement, a more quantitative demonstration will be pursued in the following section.

4.3.2 Hybrid-Entanglement of Polarization and Higher-Order LG Modes

The general state for hybrid entanglement between polarization and LG modes can be written as

$$|\Psi^{hyb}\rangle_{LG_{\pm l}} = a\,|H\rangle\,|LG_{+l}\rangle + e^{i\varphi}\cdot b\,|V\rangle\,|LG_{-l}\rangle\,, \tag{4.3}$$

where $\pm l$ stands for the order of the LG mode, i.e. the quanta of OAM per single photon. To prove the non-classicality of the bi-photon state, a similar method to the slit-wheel measurement of the last chapter is performed. Because all the spatial information is registered in one coincidence image, it is possible to discriminate between different superpositions of the LG mode by evaluating the photon number per angular region. To put it in other words, it is possible to perform the measurement of the angle-dependent photon number by masking angular regions in post-processing (see Fig. 4.7). From the measured intensity per angular region, the photon number and its error margin can be evaluated (as described in Sect. 4.2.2). Fringes that are characteristic for entanglement can be obtained for different trigger polarizations (superposition of H and V only). This is similar to the measurements with the slit wheel discussed in the previous chapter. In Fig. 4.8, fringes for hybrid-entangled states between polarization and l = ±1 and ±10 are shown. From the maxima and minima of these fringes, the visibilities can be calculated and used to test entanglement of the state. Since only measurements in two mutually unbiased bases (MUB) can be conducted, namely the superposition bases, the entanglement witness W_2 from Sect. 2.4.1 was used:

$$W_2 = vis_{D/A} + vis_{R/L} \quad \begin{cases} \leq 1 & \text{separable} \\ > 1 & \text{entangled}\,, \end{cases} \tag{4.4}$$

where, D, A, R, and L label the polarization of the trigger photon. In the following Table 4.1, all measurements to test hybrid entanglement between the polarization of the trigger photons and different LG modes of the transferred photons (l = ±1, ±2, ±3, ±5, and ±10) are presented.

Visibilities obtained from the measurements with the first-order entangled LG modes exceed the classical bound of 1 by more than 20 standard deviations. All other measurements with more complicated modal structures still violate the classical

Fig. 4.7 Schematic idea of evaluating the photon number per angular region from the measured intensity images

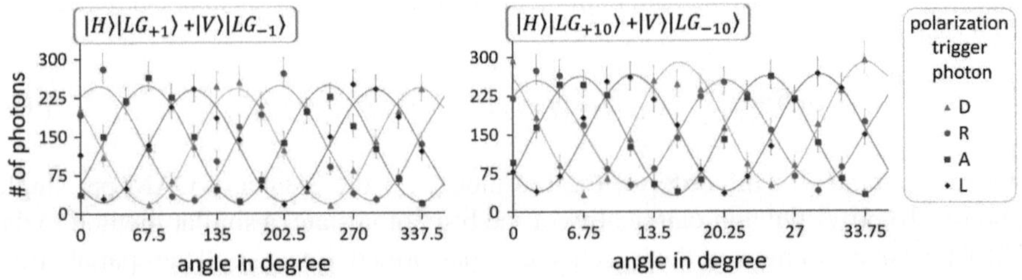

Fig. 4.8 Non-classical fringes for hybrid entanglement between polarization and LG modes with $l = \pm 1$ and $l = \pm 10$. Error bars are obtained from a Monte Carlo simulation (see Sect. 4.2.2). All angular regions in multiples of $\frac{360°}{l}$ are summed up to increase the number of photons, thereby increasing the statistical significance. This is analogous to using multiple slits

Table 4.1 Measured visibilities vis in two different MUBs and the corresponding entanglement witness W_2

State	$vis_{D/A}$	$vis_{R/L}$	W_2
$\|H\rangle \|LG_{+1}\rangle + \|V\rangle \|LG_{-1}\rangle$	0.85(2)	0.82(2)	1.68(3)
$\|H\rangle \|LG_{+2}\rangle + \|V\rangle \|LG_{-2}\rangle$	0.80(3)	0.73(4)	1.53(5)
$\|H\rangle \|LG_{+3}\rangle + \|V\rangle \|LG_{-3}\rangle$	0.75(3)	0.74(3)	1.50(5)
$\|H\rangle \|LG_{+5}\rangle + \|V\rangle \|LG_{-5}\rangle$	0.77(3)	0.73(3)	1.50(4)
$\|H\rangle \|LG_{+10}\rangle + \|V\rangle \|LG_{-10}\rangle$	0.74(4)	0.72(3)	1.46(5)

bound by around ten standard deviations. Note that no background subtraction had to be applied in any of these measurements.[6]

It is worthwhile to note that the evaluated photon numbers from the measured signal at the ICCD might be a bit smaller than they actually were in the measurement. This stems from possible saturation effects at regions of the CCD chip where many photons are registered. However, because this is only possible at the maximum of the

[6]Before the photon number per angular region was evaluated, the camera-induced readout noise mentioned earlier was subtracted. However, this subtraction should not be confused with the usual background subtraction. It is rather comparable to setting the threshold voltage in single-photon avalanche photodiode detectors, since the read-out noise is inherent to the read-out process of the ICCD. This can be adjusted as a threshold of whether a detected event is regarded as a photon or not.

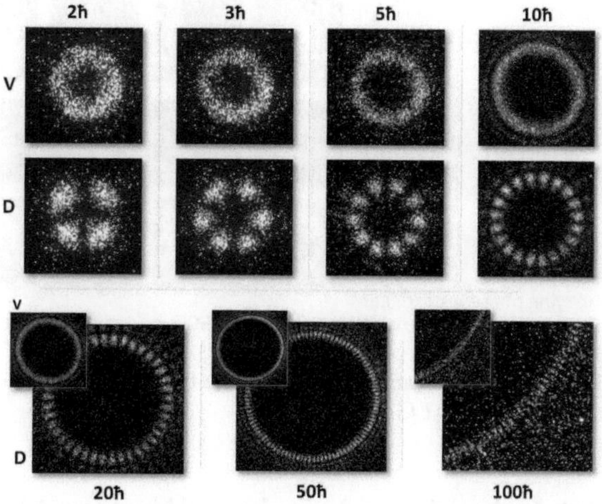

Fig. 4.9 Coincidence images of higher-order LG modes and their superposition structures. The trigger photon polarization is depicted by the *black letters*. Up to the $l = 100$th order, the structural difference of the LG mode and its superposition can be imaged

fringe, a smaller inferred photon number (than in the actual experiment) would only lead to a smaller visibility. Therefore, saturation effects would only cause smaller witness values and the presented results can be considered as a lower bound.

From the registered coincidence images that are underlying the entanglement measurements, two trigger polarizations are shown in Fig. 4.9. Vertically-polarized triggers lead to imaging of the single LG mode structure (LG_{-l}) of the transferred partner photon. Diagonal trigger polarizations lead to the expected superposition structure of $LG_{\pm l}$ modes. Additional to the measurements verifying entanglement, hybrid-entangled states with even higher-order LG modes have been generated and remotely imaged. This was done to demonstrate the advantage of the high spatial resolution of the detection method. A superposition structure can be discriminated from the ring structure of a single LG mode up to the $l = 50$th order. Even for photons that carry $100\hbar$ of OAM, a difference between the single LG mode and the superposition structure can be found. However, from those images (Fig. 4.9, where only part of the 100th-order mode is shown to illustrate the difference) it can be concluded that the resolution of the ICCD reaches its limit. Since single photons detections are spread over approximately 3 pixels on average, the resolution of the ICCD can be estimated to be 300×300 pixels. Compared to scanning or masking of single pixel detectors, this is still a very promising value although too small to resolve well a structure with 200 maxima arranged in a ring.

In Sect. 3.5.3, the LG-specific gear-like behavior was shown to appear for hybrid-entangled states. This behavior can now be imaged directly with the coincidence-imaging scheme presented above. In Fig. 4.10, the phase between the trigger polarization superposition is varied across many images and the resulting images of the LG mode structure of the entangled partner photon is shown. The gear-like behavior between the two rotations (polarizer of the trigger photon and the superposition structure) is illustrated for $LG_{\pm 1}$, $LG_{\pm 2}$ and $LG_{\pm 5}$ modes. This was shown to be advantageous in remote angular sensing applications.

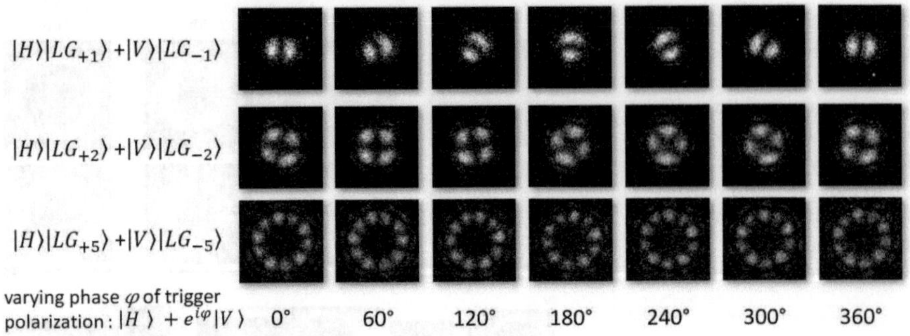

Fig. 4.10 Visualization of the gear like-behavior of hybrid-entangled states between polarization and higher-order LG modes

4.4 Coincidence-Imaging of Various Mode Families

After verifying that the measurement scheme is able to detect the single-photon structures of LG modes from an entangled hybrid bi-photon state, the next step is to test the transfer setup for different spatial modes. In Chap. 2, different mathematical coordinate systems have been discussed that lead to co-existing solutions for full sets of transverse spatial light modes. All of them can be used to encode quantum information and show features like entanglement. In the following sub-sections, hybrid-entangled states between Hermite-Gauss modes (Cartesian coordinates), Ince-Gauss modes (elliptical coordinates), and a mixture of two mode families will be shown.

Hybrid-Entanglement of Polarization and Hermite-Gauss Modes

As shown earlier in Sect. 2.2.2, Hermite-Gauss modes are solutions to the paraxial wave equation in Cartesian coordinates. To image such modes with our scheme, a hybrid-entangled state between polarization and helical Hermite-Gauss modes [10] was generated. Helical HG modes were used because a superposition of two modes with opposite handedness results in the known HG modes structures. With $hHG_{\pm}$ labeling the helical HG modes and their handedness ($\pm$), the generated state can be written as

$$|\Psi^{hyb}\rangle_{HG} = a\,|H\rangle\,|hHG_{+}\rangle + e^{i\varphi}\cdot b\,|V\rangle\,|hHG_{-}\rangle\,. \tag{4.5}$$

Triggering for different polarizations of the first photon leads to coincidence images of helical HG modes (H and V trigger) and the known Hermite-Gauss modes (superpositions of H and V). The resulting images are shown in Fig. 4.11.

Hybrid-Entanglement of Polarization and Ince-Gauss Modes

Ince-Gauss modes are a third family of transverse spatial modes that are used for creating hybrid-entangled states. IG modes are the solutions in elliptical coordinates with an additional parameter, the ellipticity ϵ. Hybrid-entangled states with IG modes

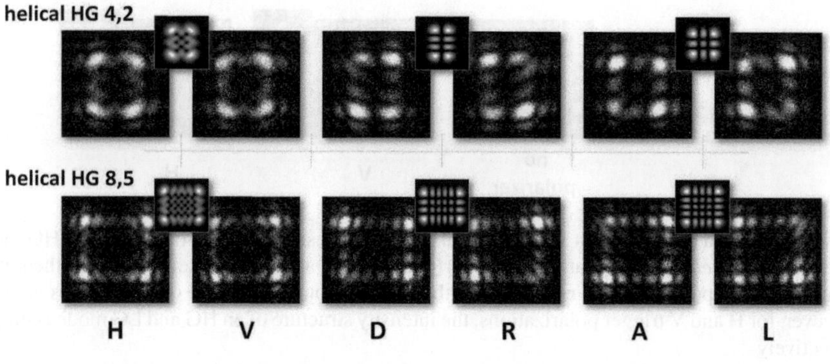

Fig. 4.11 Hybrid entanglement of helical HG modes. *Black letters* depict the polarization of the trigger photon. Small insets show the theoretical intensity structure

are written as

$$|\Psi^{hyb}\rangle_{IG^{\epsilon}} = a\,|H\rangle\,|IG_{+}\rangle + e^{i\varphi}\cdot b\,|V\rangle\,|IG_{-}\rangle\,. \tag{4.6}$$

Again, the indices $\pm$ label the handedness of the mode. Two different higher-order IG modes, each with a different ellipticity ϵ, have been entangled with the polarization of the first trigger photon. The results in Fig. 4.12 show the entangled IG modes and their superposition structures. Although the complexity of the structure increases, all characteristics like the splitting of the vortices (H and V polarized trigger) or additional nodal lines (D, R, A, and L trigger polarizations) can be recognized. This demonstrates the high resolution with which spatial features of entangled photons can be investigated with the novel coincidence-imaging scheme.

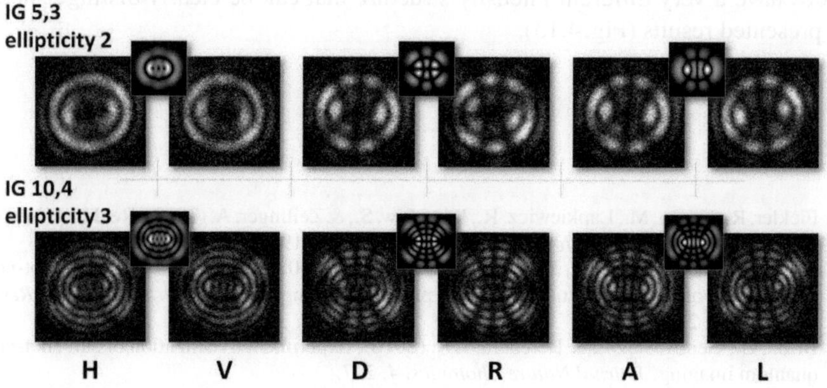

Fig. 4.12 Hybrid entanglement of IG modes. Trigger polarization is depicted by *black letters*, insets show *Black* the calculated intensity. While for H, V trigger polarizations a splitting of the vortices can be seen, additional nodal lines appear for D, R, A, L

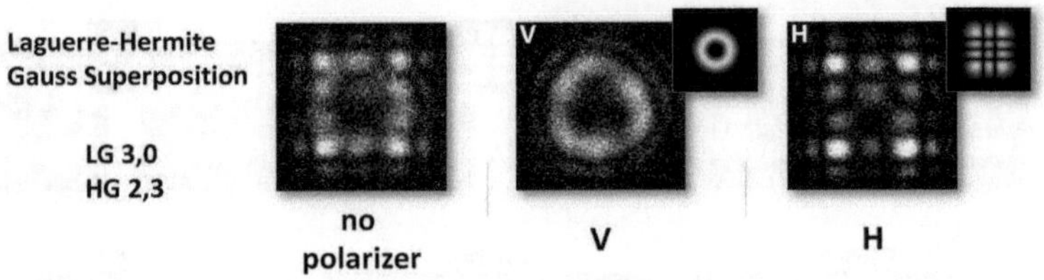

Fig. 4.13 Hybrid entanglement between polarization and a superposition of an LG and HG mode. *Black letters label* the polarizations of the trigger photon and small insets show the theoretical intensities. If no polarization is measured for the trigger photon, a mixture of both modes is found. However, for H and V trigger polarizations, the intensity structure of an HG and LG mode is imaged respectively

Hybrid-Entanglement of Polarization and Mixed Mode Families

As a last experiment and a further demonstration of the flexibility of the setup, hybrid entanglement between polarization and a mixture of two different mode families was generated and imaged. Instead of transferring the second photon to the same mode with opposite handedness, the photon is transferred to a circular LG mode (LG_{+3}) if horizontally polarized, and to a rectangular HG mode ($HG_{3,2}$) if vertically polarized. Thus, the entangled state is

$$|\Psi^{hyb}\rangle_{mix} = a\,|H\rangle\,|HG\rangle + e^{i\varphi}\cdot b\,|V\rangle\,|LG\rangle\,. \tag{4.7}$$

Now, if the trigger photon is detected with vertical polarization, the distant partner photon is remotely projected into an LG mode that carries OAM. If the trigger photon is measured to be in horizontal polarization, the transferred photon is in an HG mode with no angular momentum. Thus, before any measurement takes place, the transferred photon is in a superposition of an LG mode and an HG mode. Both modes have a very different intensity structure that can be clearly distinguished in the presented results (Fig. 4.13).

References

1. Fickler, R., Krenn, M., Lapkiewicz, R., Ramelow, S., & Zeilinger, A. (2013). Real-time imaging of quantum entanglement. *Journal Scientific Reports*, *3*, 1914.
2. Blanchet, J.-L., Devaux, F., Furfaro, L., & Lantz, E. (2008). Measurement of sub-shot-noise correlations of spatial fluctuations in the photon-counting regime. *Journal Physical Review Letters*, *101*, 233–604.
3. Brida, G., Genovese, M., & Berchera, I. R. (2010). Experimental realization of sub-shot-noise quantum imaging. *Journal Nature Photonics*, *4*, 227.
4. Moreau, P. A., Mougin-Sisini, J., Devaux, F., & Lantz, E. (2012). Realization of the purely spatial Einstein-Podolsky-Rosen paradox in full-field images of spontaneous parametric down-conversion. *Journal Physical Review A*, *86*, 010–101.
5. Edgar, M. P., et al. (2012). Imaging high-dimensional spatial entanglement with a camera. *Journal Nature Communications*, *3*, 984.

6. Jost, B. M., Sergienko, A. V., Abouraddy, A. F., Saleh, B. E. A., & Teich, M. C. (1998). Spatial correlations of spontaneously down-converted photon pairs detected with a single-photon-sensitive CCD camera. *Journal Optics Express*, *3*, 81.
7. Abouraddy, A. F., Nasr, M. B., Saleh, B. E. A., Sergienko, A. V., & Teich, M. C. (2001). Demonstration of the complementarity of one- and two-photon interference. *Journal Physical Review A*, *63*, 063–803.
8. Jedrkiewicz, O., et al. (2004). Detection of sub-shot-noise spatial correlation in high-gain parametric down conversion. *Journal Physical Review Letters*, *93*, 243–601.
9. Pires, H. D. L., Monken, C. H., & van Exter, M. P. (2009). Direct measurement of transverse-mode entanglement in two-photon states. *Journal Physical Review A*, *80*, 022–307.
10. López-Mariscal, C., & Gutiérrez-Vega, J. C. (2007) Propagation dynamics of helical Hermite-Gaussian beams. In *Proceedings of the SPIE*, (Vol. 6663, pp. 666–307).

Chapter 5
Entanglement of Complex Photon Polarization Patterns in Vector Beams

5.1 Preamble

The content of the following chapter is based on the publication "Quantum Entanglement of Complex Photon Polarization Patterns in Vector Beams" [1].

In the last chapter a method was developed to use coincidence images to evaluate the photon distributions spatially with high resolution and demonstrate effects of entanglement. Earlier, the transfer setup was introduced, with which it is possible to transfer polarization entanglement to more complex bi-photon states, such as high OAM or hybrid-entangled photon pairs. Now, the complexity of the generated bi-photon state can be increased even further by including the vectorial property of light and combining it in a non-trivial way with transverse spatial modes, i.e. superpositions of different, orthogonally-polarized light modes, which leads to varying polarization across the beam extent.

As mentioned in Chap. 3, those complex structures can enhance the atom-photon coupling efficiency up to 100 % [2], which suggests that entangled vector photons will facilitate entangling separated atoms in broader quantum networking tasks. Another Interesting application of such complex states might be cluster-state quantum computing, where the polarization and spatial degree of freedom increases the amount of encoded qubits per photon [3]. The entanglement of such states could also be used for teleportation of complex structures, which might be advantageous for both applied and foundational quantum experiments. However, all described future prospects require an efficient way to generate, entangle, and investigate the complex photonic states. With the methods introduced and established in the previous chapters, it is possible to achieve such tasks.

R. Fickler, *Quantum Entanglement of Complex Structures of Photons*,
Springer Theses, DOI 10.1007/978-3-319-22231-8_5

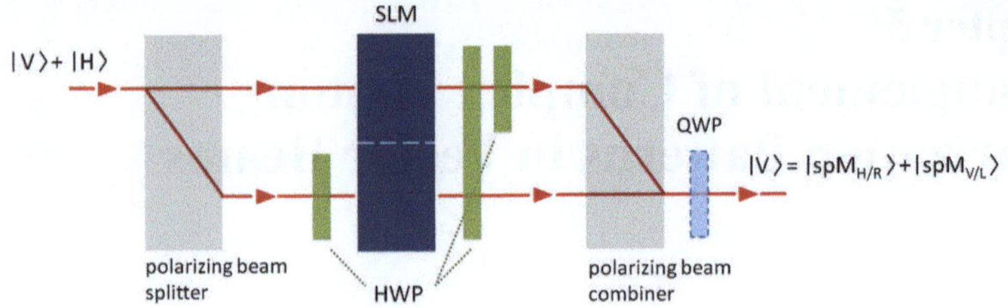

Fig. 5.1 Simplified sketch of the setup for generating vector beams. The difference from the transfer setup introduced earlier is that the which-path information is not erased by a polarizer after the interferometer. Instead of the polarizer, a quarter wave plate (QWP) might be placed behind the interferometer if a superposition of left- (L) and right-handed (R) circularly-polarized light is required. Note that the actual setup was realized again as a folded Sagnac-like interferometer, as shown in Fig. 3.7

5.1.1 Experimental Generation of Vector Beams

In Sect. 2.2.5, different types of vector beam families were introduced theoretically. The general state of such vector beams can be written as a superposition of two different, orthogonally-polarized transverse spatial modes. The general state from Eq. 2.23 can be written in a simplified version as

$$|Vec\rangle = a\,|spM_{H/R}\rangle + e^{i\varphi}\cdot b\,|spM_{V/L}\rangle\,, \tag{5.1}$$

where a, b, and φ are real and $a^2+b^2=1$. $spM_{X/Y}$ labels the spatial modes and its index denotes the polarization (nomenclature as introduced earlier). Experimentally, this superposition state has been realized in many different ways, e.g. with segmented wave plates [4], interferometric setups [5] or by using a so-called q-plate[1] [7], to name a few (more can be found in [8]). The transfer setup introduced previously is also able to create vector beams by simply removing the polarizer after the interferometer (replacing it with a quarter wave plate if R or L polarization is required; see Fig. 5.1). The main advantage of using the interferometric setup is its enormous flexibility in generating different vector beams, which was already demonstrated in [9, 10]. However, the phase stability of such a setup is crucial (especially at the single-photon level), thus the folded Sagnac-like structure is ideal.

5.1.2 Measurement of the Polarization State of Light

In Sect. 2.2.3, different polarizations have been discussed and the Stokes parameters have been introduced (Eq. 2.18). The latter can be directly related to intensity mea-

[1]q-plates are liquid crystal devices that transfer a photon's spin to its orbital angular momentum. More details can be found in [6].

surements along six polarization directions (H, V, D, A, R, and L). Thus, in order to determine experimentally the polarization of a light field, six intensity measurements have to be conducted. The whole procedure is called polarization tomography. The relations between the intensities I, the Stokes parameters S, and the amplitudes u and phases δ of the polarization are

$$S_0 = I_H + I_V = u_x^2 + u_y^2 \tag{5.2a}$$
$$S_1 = I_H - I_V = u_x^2 - u_y^2 \tag{5.2b}$$
$$S_2 = I_D - I_A = 2u_x u_y \cos\delta \tag{5.2c}$$
$$S_3 = I_R - I_L = 2u_x u_y \sin\delta. \tag{5.2d}$$

The intensities of horizontal and vertical polarization measurements reveal the amplitudes in the x and y directions, i.e. $I_H = u_x^2$ and $I_V = u_y^2$. The phase δ can be uniquely determined from the remaining four intensity measurements. As shown in Eq. (2.19), the parameters describing the polarization ellipse, i.e. its ellipticity ϵ and orientation θ, can be reconstructed with the measured Stokes parameters as well.

5.2 Hybrid-Entanglement of Polarization and Vector Beams

Similar to the setup in the previous chapter, polarization entanglement is created first using the Sagnac source described earlier (see Fig. 5.2).[2] In a second step, one of the two photons is coupled to a single-mode fiber for Gauss mode filtering, delayed by 35 m, and brought to the adapted transfer setup (see Fig. 5.2a). Then, the photon is coherently converted into a vector-polarization mode (herein called "vector photon") by interferometrically superposing two different spatial modes with orthogonal polarizations. The resulting hybrid-entangled two-photon state can be written as

$$|\Psi^{Vec}\rangle = a\,|H\rangle\,|spM_{V/R}\rangle - e^{i\varphi}b\,|V\rangle\,|spM_{H/L}\rangle\,, \tag{5.3}$$

where a, b, and φ are real and $a^2 + b^2 = 1$, H and V denote the horizontal and vertical polarization of the unchanged photon, $spM_{X/Y}$ and its index represents the different spatial modes and their polarizations and the positions of the ket-vectors label the different photons.

To demonstrate the generated hybrid entanglement, polarization-sensitive coincidence-imaging is performed (see Fig. 5.2b). Analogous to the scheme in the last chapter, a polarization measurement of the unchanged photon is performed with a QWP, a polarizer, and a single-photon detector (avalanche photodiode). The detec-

[2]The source was adjusted such that the bi-photon polarization state was in the $|\Psi^-\rangle$ state (see Eq. (2.38)). As already mentioned in Chap. 2, this state is special, since it is the only Bell-state that is anti-correlated in all three MUBs. This means, that if the first photon is found in a certain polarization state the second one will be orthogonally polarized.

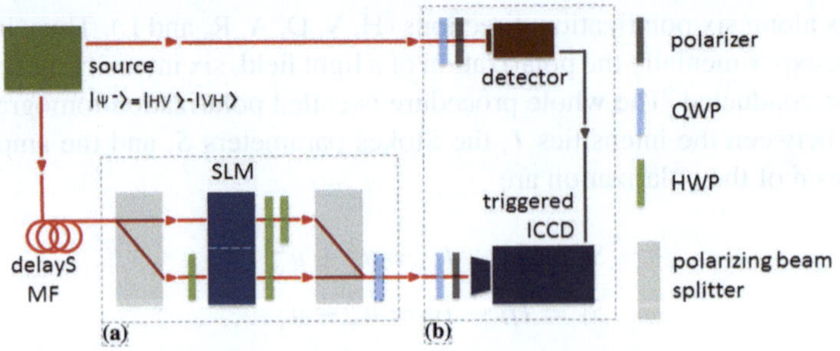

Fig. 5.2 Sketch of the experimental setup. Polarization entanglement between two photons is created (*green box*). **a** One photon is transferred to a vector photon by the adapted transfer setup. **b** The generated hybrid entanglement is analyzed by measuring the polarization of the unchanged photon in combination with transverse polarization tomography of the transferred photon (see main text for more information)

tion signal is used to trigger the intensified CCD camera. An additional QWP and polarizer are inserted in front of the ICCD camera to measure spatially the polarization distribution of the incident photons and thereby realize polarization-sensitive coincidence-imaging. For each trigger polarization, spatially-resolved polarization tomography of the vector photon is performed to reconstruct its complex polarization pattern (see next section).

For the following measurements, a different ICCD camera to the one described in the previous chapter was used. In this camera, a third generation image intensifier is implemented (Andor iStar A-DH334T-18U-73) that increases the performance of the coincidence scheme. Most of the characteristics, like the resolution of the CCD chip, the MCP, and the repetition rate of the gating (max. 500 kHz) are the same. In addition, the detection efficiency for 810 nm photons and the speed of the gating window are better. The quantum efficiency specified by the manufacturer is 20 % at 810 nm,[3] which increases the performance of the ICCD detection by a factor of six. This enhancement comes with a rise of the dark counts (approximately 3 times more). Although the minimal gating window is specified to be 2 ns, the following experiments were performed with a window of 5 ns time duration because of a better detection efficiency (intensifier transmission efficiency: 30 % at 2 ns and >90 % at 5 ns). Taking all described factors into account, the ICCD camera used in the following experiments has 3 times better signal-to-noise ratio than the test-camera of the previous chapter.

The different type of intensifier leads to a different registered signal count per single photon. Hence, the characterization described in Sect. 4.2.2 was performed again. More than 6600 single-photon detections were analyzed and an average signal of 350 counts per single photon was found. As before, the measured probability distribution was used to simulate the error margin of each photon number with a

[3] Analogous to its definition in Sect. 4.1, the quantum efficiency of an ICCD can be regarded as a measure of its electrical sensitivity to a photon.

Monte Carlo algorithm. In contrast to the ICCD camera used before, the standard deviation of the simulation was found to be comparable to the one expected from Poissonian count statistics. Thus, the latter one was used for all error calculation. After the characterization of the ICCD camera, it can be used to image the complex spatial structures of the entangled vector photons.

5.2.1 Transverse Polarization Tomography Using Coincidence Images

To reconstruct the complex polarization pattern, local polarization tomography (see Sect. 5.1.2) was performed in the transverse spatial domain with the help of a QWP and polarizer in front of the ICCD camera. For every trigger polarization, six settings of the polarization projection in front of the ICCD camera were used. For every setting, an exposure time of 15 s was used to image the modal structure by accumulating many single-photon events. Although shorter measurement times would have led to very good results, the long exposure time was used to get better statistics for later evaluation of the photon numbers within small transverse regions. The source was adjusted such that after the polarization measurement and the fiber coupling of the trigger photon the rate of the detector was around 500 kHz (maximal gating rate of the ICCD). Some example reconstructions using coincidence images can be found in Figs. 5.3 and 5.4.

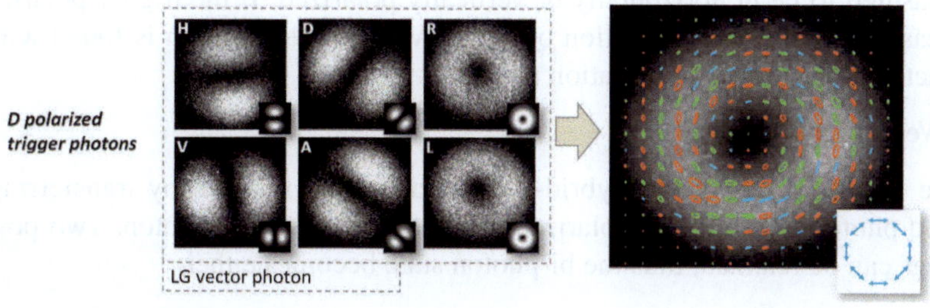

Fig. 5.3 Example of transverse polarization tomography using coincidence images. The presented measurements were done with diagonally-polarized trigger photons and vector photons composed of circularly-polarized, first-order LG modes $l = +1$ and $l = -1$. *Left* The recorded intensity distribution depends on the polarization settings in front of the ICCD camera (*white letters*). *Right* Local polarization tomography, where 4×4 pixel arrays were used to reconstruct the pattern. Every 5th polarization vector is shown. The colors depict the type and handedness of polarization (*blue* = linear, *red* = right-elliptic, *green* = left-elliptic), its size corresponds to the detected intensity. The polarization vector was regarded as linear if the ratio between the major and minor axes of the polarization ellipse was bigger than 10. The average intensity distribution is overlaid with the reconstructed polarization pattern. Both, the recorded coincidence images and the reconstructed azimuthal pattern, fit well to the theoretical predictions (small insets)

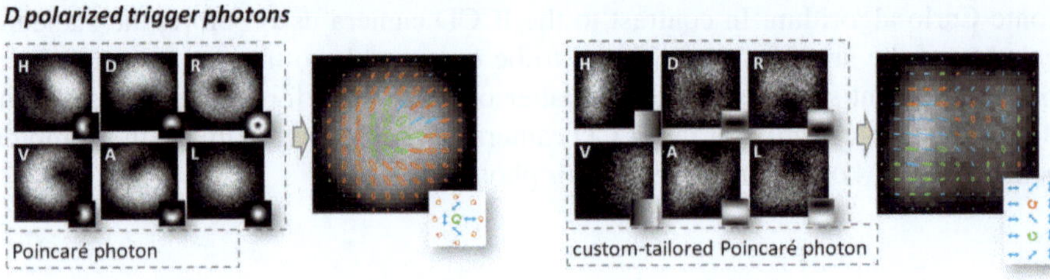

Fig. 5.4 Examples of transverse polarization tomography from coincidence images with Poincaré photons (photons exhibiting the polarization pattern of Poincaré beams). Again, all measurements were done with D polarized trigger photons. Arrangement, insets, labeling, and color coding as in Fig. 5.3. Here, each vector is calculated from a 10 × 10 pixel array and every third is shown. *Left* Poincaré photon built by a right-handed circularly-polarized first-order LG mode and left-handed circularly-polarized Gauss mode. *Right* Custom-tailored Poincaré photon

5.2.2 Remote Preparation of Single-Photon Vector Beams

In the last section, only the diagonal trigger polarization was used, and the pattern was reconstructed from the recorded coincidence images for different settings of the polarization analyzer (QWP and polarizer). With the same procedure but different trigger polarizations, various vector photons can be remotely generated and their patterns reconstructed. In the following section, different types of cylindrical vector photons (HG and LG types), as well as different Poincaré photons, are investigated. Their common feature is the uniformly distributed polarization, seen when the trigger is measured to be in horizontally or vertically polarized. If the trigger polarization is measured to be a superposition of H and V, the vector photon is found with the characteristic complex polarization pattern.

HG Vector Photons

In the first experiment, the hybrid-entangled state is realized by transferring the second photon to a linearly-polarized Hermite-Gauss vector photon. Two possible settings can be realized, thus the bi-photon state becomes either

$$|\Psi_{HG1}^{Vec}\rangle = a\,|H\rangle\,|HG_{1,0/V}\rangle - e^{i\varphi}b\,|V\rangle\,|HG_{0,1/H}\rangle \tag{5.4}$$

or

$$|\Psi_{HG2}^{Vec}\rangle = a\,|H\rangle\,|HG_{0,1/V}\rangle - e^{i\varphi}b\,|V\rangle\,|HG_{1,0/H}\rangle\,, \tag{5.5}$$

where the indices label the order (two numbers) and polarization (letter) of the HG mode. All reconstructed patterns can be seen in Fig. 5.5. The cylindrically symmetric structures of the azimuthal and radial polarization can be recognized for A-polarized trigger photons of the state $|\Psi_{HG1}^{Vec}\rangle$ (Fig. 5.5 upper row) or the state $|\Psi_{HG2}^{Vec}\rangle$ (Fig. 5.5 lower row). Additionally, the so-called anti-vortex pattern, where the linear polarization vector rotates contrary to its angular position (Fig. 5.5 D-trigger) can be seen after reconstruction.

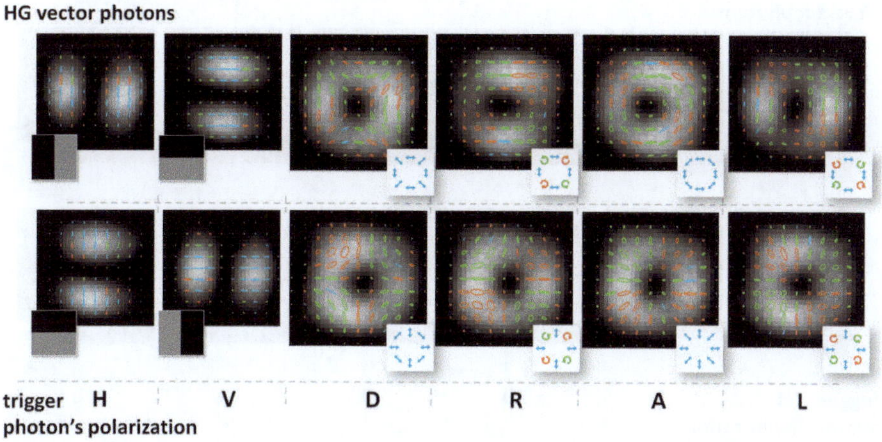

Fig. 5.5 Measured polarization patterns for different HG vector photons entangled with the polarization of their partner photons. As before, the average registered intensity distribution for each trigger polarization (*black letters*) is shown and overlaid with the measured polarization pattern (the polarization of every third group of pixels is shown). Colors depict the type and handedness of the measured polarization (coding as in Fig. 5.3). The imprinted phases are shown in *gray scale* insets ($0 =$ *black* / $\pi =$ *gray*). *Upper row* HG vector photons from Eq. (5.4). *Lower row* HG vector photons from Eq. (5.5)

LG Vector Photons

In a second experiment, vector photons built from Laguerre-Gauss modes are generated and entangled with the polarization of the trigger photon. The vector photons are built from first-order LG modes with opposite handedness of the helical phase front ($\pm l$) and of the circular polarization (R, L). Hence, the bi-photon state can be expressed as

$$|\Psi_{LG1}^{Vec}\rangle = a\,|H\rangle\,|LG_{1,0/R}\rangle - e^{i\varphi}b\,|V\rangle\,|LG_{-1,0/L}\rangle \tag{5.6}$$

or

$$|\Psi_{LG2}^{Vec}\rangle = a\,|H\rangle\,|LG_{-1,0/R}\rangle - e^{i\varphi}b\,|V\rangle\,|LG_{1,0/L}\rangle\,, \tag{5.7}$$

with analogous labeling as for the HG vector photons. All measured polarization patterns can be found in Fig. 5.6. If the trigger photon is detected in H or V polarization, the transferred photon is registered in an LG mode with circular polarization (after the transfer setup, the linear polarizations are changed to circular with the additional QWP). For superpositions of H and V polarization, the vector photon's varying polarization pattern appears. Due to entanglement, different polarizations of the trigger photon result in different polarization patterns of its partner vector photon. Thus, the Bloch sphere for vector photons–the higher-order Poincaré sphere–can be imaged [11, 12]. For LG vector photons, only linear polarizations are found. In Fig. 5.6, the upper row shows the cylindrically symmetric modes denoted by Eq. (5.6) in an azimuthal (D trigger), spiral (L or R trigger), and radial (A trigger) configuration. In

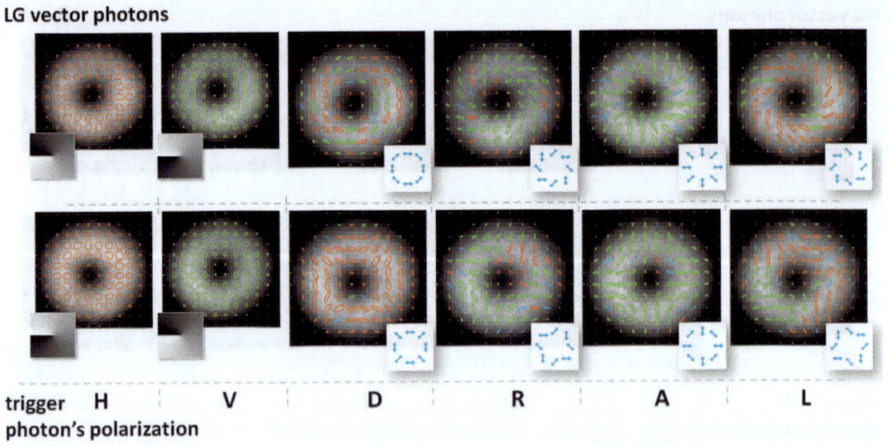

Fig. 5.6 Measured polarization patterns for different LG vector photons entangled with the polarization of their partner photons (the trigger polarization is denoted by *black letters*). Color coding and layout analogous to the previous Fig. 5.5. The imprinted phases for H and V polarization triggers are presented by gray scale insets (linear from $0 =$ *black* to $2\pi =$ *white*). *Upper row* LG vector photons from Eq. (5.6). *Lower row* LG vector photons from Eq. (5.7)

the lower row, the polarization patterns of the state $|\Psi_{LG2}^{Vec}\rangle$ have been measured. All superpositions show the anti-vortex structure, which rotates with the change of the trigger polarization.

Poincaré Photons

The class of Poincaré beams is tested in a third experiment. This class of light modes contains polarization singularities and has been realized for the first time at the single-photon level. The bipartite state, after one photon has been transferred, can be written as

$$|\Psi_{P1}^{Vec}\rangle = a\,|H\rangle\,|LG_{1,0/R}\rangle - e^{i\varphi}b\,|V\rangle\,|LG_{0,0/L}\rangle \tag{5.8}$$

or

$$|\Psi_{P2}^{Vec}\rangle = a\,|H\rangle\,|LG_{0,0/R}\rangle - e^{i\varphi}b\,|V\rangle\,|LG_{1,0/L}\rangle\,, \tag{5.9}$$

where the Poincaré photon is realized by a superposition of a first-order LG mode and a fundamental Gauss mode ($LG_{0,0}$), each with the opposite circular polarization. The two hybrid-entangled states $|\Psi_{P1}^{Vec}\rangle$ and $|\Psi_{P1}^{Vec}\rangle$ that have been generated in this experiment result in the patterns displayed in Fig. 5.7 for various trigger polarizations. The characteristic lemon (upper row) and star (lower row) configurations [13] can be observed for triggers of D, A, R, and L polarization. Poincaré photons exhibit two interesting polarization singularities: C-point singularities, which are points where the orientation of the polarization ellipse is undefined, and L-line singularities, where the handedness of the polarization ellipse is undefined. Both types have not been detected in quantum optical experiments before. In Fig. 5.7, the gray dotted circles

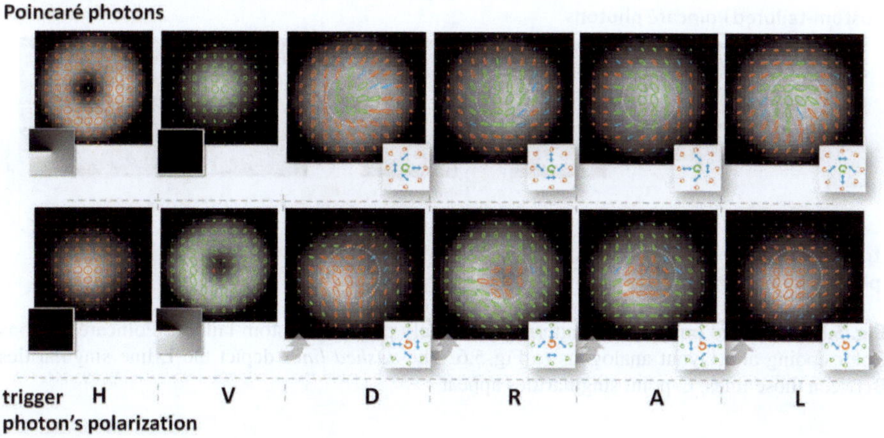

Fig. 5.7 Measured polarization patterns for hybrid-entangled pairs of Poincaré photons and uniformly-polarized partner photons. Color coding and layout is analogous to Fig. 5.6. *Upper row* Poincaré photons in a so-called lemon configuration (state 5.6). *Lower row* Poincaré photons with the star configuration (state 5.7). Here, the patterns were found to be inconsistent with the theory that predicts the reconstructed patterns for "neighboring" polarization triggers (depicted with the *gray arrows*). The reason was most likely a phase change due to misalignments in the transfer setup (see main text). The *dashed circles* depict the L-line singularities and the centers exhibit a C-Point singularity

depict the approximate position of the L-line singularity, while the center of each Poincaré photon exhibits a C-point singularity. Hence, it was possible to demonstrate that single photons can have polarization singularities and that photon pairs can be entangled in such highly complex patterns. While the first row in Fig. 5.7 shows a reasonable agreement with theoretical prediction, the second row does not. All modes are rotated by 90° with respect to the pattern theoretically predicted. A closer investigation reveals that a phase change between the two constructing modes lead to a rotation of the polarization pattern. Therefore, it is most likely that the measurements presented in the lower row of Fig. 5.7 disagree with the theoretical predictions due to misalignments of the transfer setup. A very small path difference, i.e. a fraction of a wavelength, is translated to a phase change, which changes the generated polarization pattern (here it is rotated).

Custom-Tailored Poincaré Photons

In the last experiment, the flexibility of the transfer setup was used to create custom-tailored Poincaré photons that are entangled with the polarization of the trigger photon. As described theoretically in Sect. 2.2.5, the modes are built by a superposition of two square shaped intensity structures, where one is horizontally polarized while the other one has a vertical polarization. For each mode, the intensity changes linearly from left to right and the phase varies linearly from top to bottom. To experimentally generate such modes the phase modulation of the SLM has to be extended to modulate the intensity as well. The idea is to image the SLM surface on the ICCD

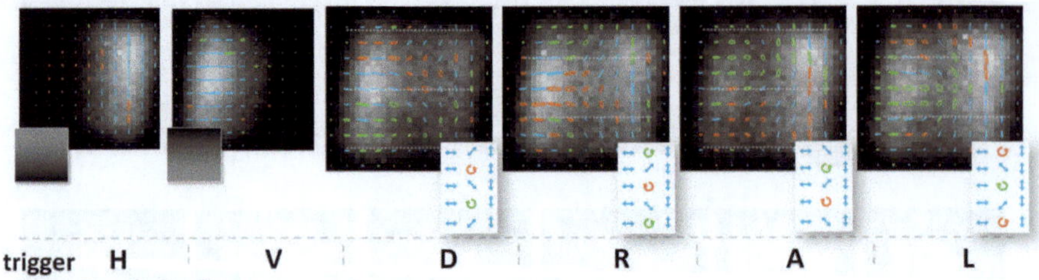

Fig. 5.8 Measured polarization patterns of remotely prepared custom-tailored Poincaré photons. Color coding and layout analogous to Fig. 5.6. The *dashed lines* depict the L-line singularities. Between those lines, C-point singularities appear

camera chip and adjust the diffraction efficiency of an additionally displayed grating according to the desired intensity structure. The holographic pattern, consisting of a phase pattern plus a phase grating, is thereby able to modulate the phase and the intensity of the photons simultaneously.[4] With this technique, any custom-tailored polarization pattern of the vector photon can be created and entangled with its partner photon's polarization. Hence, the bi-photon state is

$$|\Psi_{ctP}^{Vec}\rangle = a\,|H\rangle\,|\underleftarrow{SpM}\downarrow_V\rangle - e^{i\varphi} b\,|V\rangle\,|\underrightarrow{SpM}\uparrow_H\rangle\,, \tag{5.10}$$

where the horizontal arrows depict the change in intensity and the vertical arrows the phase gradient. Both are pointing in direction of decrease. H and V label the polarization of the square-shaped mode. The measured polarization patterns can be found in Fig. 5.8. As described earlier in Sect. 2.2.5, the polarization patterns continuously changes from H (left side) to V (right side) through all possible polarizations as a function of the vertical position. Similar to the previous Poincaré photons, C-points and L-lines can be found in the polarization patterns of the custom-tailored Poincaré photons.

5.3 Entanglement of Complex Polarization Patterns in Vector Photons

So far, the high-contrast intensity images for different trigger polarizations and the subsequent changes of the polarization patterns have indicated the successful generation of entanglement. However, to demonstrate the non-classicality of the state

[4]The blazed phase grating is simply added to the targeted phase pattern (modulo 2π), which leads to a phase hologram. Since only the first diffraction order is imaged to the ICCD chip, adjusting the diffraction efficiency leads to a modulation of the intensity. For more information about this technique the reader is referred to [14].

quantitatively, a local evaluation of three different types of entanglement measures in the polarization DOF was performed for the hybrid-entangled photon pairs. The change in the appearing entanglement pattern was investigated while the polarization settings were altered. In all the following measurements, coincidence images for appropriate polarization combinations (trigger and image polarization) were registered. The average photon number within regions of 10×10 pixels was evaluated by the method described in the previous sections. From these local measurements, the value for each entanglement criterion was calculated. Note that this is only feasible due to the fast coincidence-imaging scheme. Relying on counting single-photon events in sparse images or scanning single-pixel detectors across the beam would have been extremely challenging and time consuming, especially for the later observation of the setting-dependent entanglement pattern.

5.3.1 *Direct Visualization of Different Strength of Entanglement Tests*

At first, three different types of entanglement tests that were introduced in Sect. 2.4 are performed. A brief recapitulation of all entanglement measures might be useful to better understand the presented measurement results. The first measure of entanglement is the entanglement witness employing measurements in 3 mutually unbiased bases from Sect. 2.4.1. It fully relies on quantum mechanical predictions

$$W_3 = |\hat{\sigma}_x \otimes \hat{\sigma}_x| + |\hat{\sigma}_y \otimes \hat{\sigma}_y| + |\hat{\sigma}_z \otimes \hat{\sigma}_z|, \tag{5.11}$$

where $\hat{\sigma}_x$, $\hat{\sigma}_y$, and $\hat{\sigma}_z$ denote the single-qubit Pauli matrices for the bipartite systems (positions label the two photons). The witness is bounded by 1 for any separable state. Thus, values larger than 1 verify entanglement. As a second criterion, the steering inequality from Sect. 2.4.2 is calculated from the same measurements

$$S_{St} = |\hat{\sigma}_x \otimes \hat{\sigma}_x|^2 + |\hat{\sigma}_y \otimes \hat{\sigma}_y|^2 + |\hat{\sigma}_z \otimes \hat{\sigma}_z|^2. \tag{5.12}$$

If measurement results exceed the classical bound of 1, the non-classicality of the state, i.e. "non-local steering of the vector photon" is proven. Because weaker assumptions are made than for the witness, the steering inequality is violated by a smaller class of states. The last test for entanglement is the Bell-CHSH-inequality (Sect. 2.4.3)

$$S = |E(\alpha, \beta) - E(\alpha', \beta) + E(\alpha, \beta') + E(\alpha', \beta')|, \tag{5.13}$$

where α, α', β, and β' denote different measurement settings (orientations of the polarizer) and E is the normalized expectation value for photon pairs to be detected with these settings. The Bell-CHSH-value S is bounded by the upper limit of 2 for all local realistic theories (definitions, discussion, and calculation can be seen in

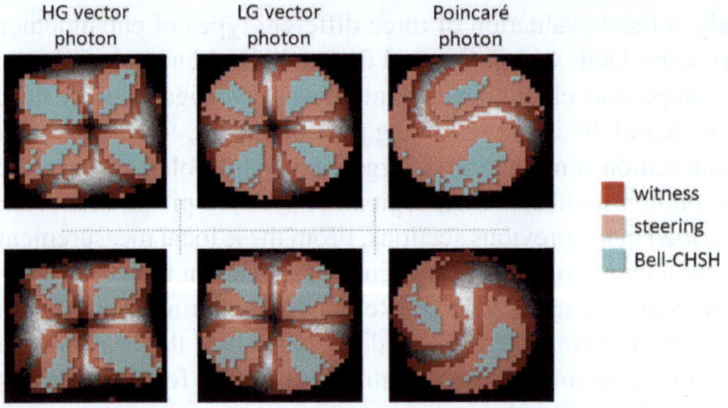

Fig. 5.9 Polarization entanglement patterns of different vector photons. Three different tests for polarization entanglement were performed: witness (*dark red*), steering (*light red*), and Bell-CHSH (*blue*). The colors depict regions that were successfully tested, i.e. regions where more than 100 photons contributed to every term of the criterion and the classical bound is exceeded by at least 3 standard deviations. Upper and lower measurements correspond to the upper and lower vector photon modes from the previous Sect. 5.2.2

Sect. 2.4.3). Therefore, a violation of the Bell-CHSH-bound excludes an even larger class of states, since it is harder to violate and it does not rely on any quantum mechanical assumptions. In Fig. 5.9, regions are shown where the witness, the steering, or the Bell-CHSH-inequalities prove entanglement. For LG vector photons and Poincaré photons, the Bell-CHSH-inequality was tested with L, R, D, and A polarization triggers instead of the usual H, V, D, and A polarization settings. This was done due to the nature of their polarization pattern. In addition, the Bell-CHSH-inequality was measured for perfect correlation and perfect anti-correlation of the polarization of the photons. Otherwise, only half of the blue regions would appear. This stems from the mathematical structure of the Bell-CHSH-inequality, which is sensitive to the type of correlation, unlike the other two entanglement measures. The observed patterns depict visually that the weaker the assumptions of the criterion, the smaller the regions of successful entanglement demonstration. Note that the "yin-yang" like entanglement pattern of the Poincaré photons arises due to slight misalignments and is already indicated in the actual coincidence images that can be seen in Fig. 5.4.

Apart from the visual demonstrations of the entanglement tests and the quantitative proof of the non-classicality of the state, an interesting observation can be made. Only a part of the transverse beam profile shows entanglement in the polarization degree of freedom. This interesting feature will be investigated in more detail in the following section.

5.3.2 Comparison of Entanglement Patterns in Different Vector Photons

Since the polarization is varying across the transverse beam and only small regions show polarization entanglement, a natural question arises. Is the observed entanglement pattern dependent on the polarization setting? For a radially-polarized beam, one can easily imagine that the measurement settings have to be adjusted according to the rotation of the polarization vector to show entanglement. To phrase it in other words, does the entanglement pattern only appear because the polarization measurement settings were not adjusted accordingly?

To answer the posed question and to account for the varying polarization, different "reference frames", i.e. polarizer angles, were used. With each setting (or reference frame), a Bell-CHSH-test was performed. The Bell-CHSH-criterion was utilized, since it is the strongest.

In Fig. 5.10, the polarization entanglement of LG vector photons is investigated. By altering the reference frame of the measurement settings in front of the ICCD and performing the Bell-CHSH-test, the structure of proven entanglement is found to rotate as well. If all regions of a successful Bell-CHSH-test are summed up, the whole beam shows non-classical features. Hence the LG vector photon is entangled in polarization with the trigger photon independent of the transverse region.

So far, the observed entanglement patterns of LG and HG vector photons were found to be very similar (see Fig. 5.9). However, if the entanglement is tested in rotated reference frames with HG vector photons, a substantial difference between the two vector photons becomes apparent. Here, the regions of successful tests of the

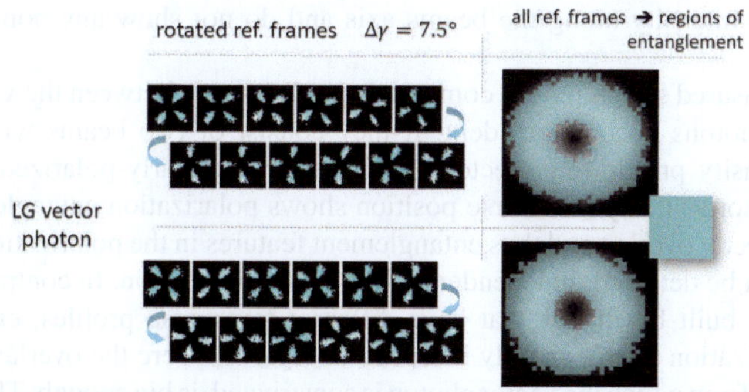

Fig. 5.10 Entanglement pattern of LG vector photons for different orientations (rotated reference frames) of the polarizer in front of the ICCD. In the *blue regions*, the Bell-CHSH-inequality was violated by at least three standard deviations and more than 100 photons contributed to each term of the inequality. *Left* The *blue* structure rotates with the change of the reference frame. *Right* Summing over all tested reference frames reveals the full polarization entanglement structure. The inset on the *right* shows the theoretical expectation. Here, no experimental restriction on a minimum photon number contribution is necessary and thus entanglement is found even in the center and outside of the beam

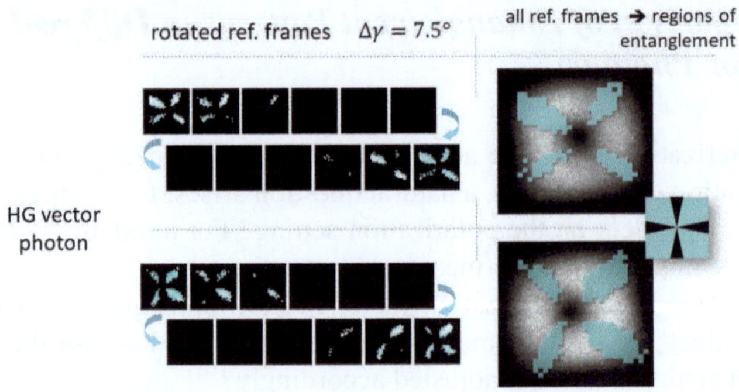

Fig. 5.11 Entanglement pattern of HG vector photons for different orientations (rotated reference frames) of the polarizer. Color coding and layout is analogous to Fig. 5.10. *Left* The *blue regions* get smaller, vanish, and reappear as the reference frame is rotated. *Right* The sum over all tested reference frames reveals the full polarization entanglement pattern where the photons are only entangled in polarization if found in certain transverse regions

Bell-CHSH entanglement measure get smaller, vanish, and reappear as the setting is rotated by 180°. The sum over all reference frames shows a structure where the photon pair is only entangled in some regions, while it is always separable between them (Fig. 5.11).

A similar feature can be observed for hybrid-entangled pairs, where the vector photon is encoded as a Poincaré photon. Although the angular position of the pattern changes with the rotation of the reference frame, no entanglement is found in the center of the beam. Compared to LG vector photons, where no entanglement was found in the center of the beam due to the small photon number, Poincaré photons have a high intensity along the beams axis and do not show any non-classicality there.

If all measured structures are compared, the difference between the various types of vector photons becomes evident: if they consist of two beams with the same spatial intensity profile (e.g. vector photons from circularly-polarized LG modes of the same order), any transverse position shows polarization entanglement. Both modes perfectly overlap and thus entanglement features in the polarization degree of freedom can be detected, independent of the transverse region. In contrast, if vector photons are built by modes that have different transverse profiles, entanglement in the polarization DOF can only be found in regions where the overlap of the two amplitudes, from which the vector photon is constructed, is big enough. Theoretically, only very small regions are separable, because a small overlap between the two modes already leads to non-maximally entangled photon pairs. However, in the experiment, the modal overlap needs to be bigger due to experimental imperfections that make it harder to reveal features of polarization entanglement. This can be seen for example in the entanglement pattern of the last Poincaré photon, where slight misalignments lead to a pattern that does not fit very well to the theoretically expected one.

Nevertheless, a novel feature of certain entangled vector photons is found. They can be both entangled and not entangled in polarization at the same time, dependent on the transverse spatial region where they are detected. This interesting feature can be explained intuitively: at transverse positions where a polarization measurement can reveal the mode of the photon (and thus path of the photon inside the interferometer of the transfer setup), there is no superposition of spatial modes. Hence, no entanglement is measurable. On the contrary, if the polarization measurement cannot give any information about the mode of the photon (and the path in the transfer setup), polarization entanglement is found (Fig. 5.12).

Of course, all discussed features hold for custom-tailored Poincaré photons as well. Due to the non-perfect intensity modulation during the process of generation in the transfer setup, the modal overlap is too small to allow for broad regions of a violation of the Bell-CHSH-inequality (see Fig. 5.13). However, a change of the trigger polarizations used to test entanglement show that the entanglement patterns shift according to the theoretical calculation. From this observation, it can be deduced

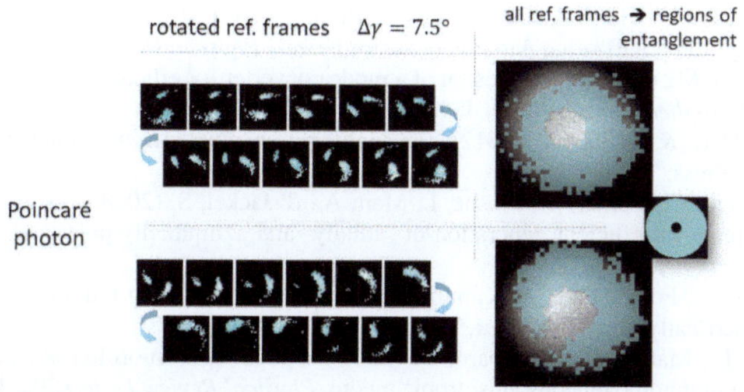

Fig. 5.12 Entanglement pattern of Poincaré vector photons for different orientations (rotated reference frames) of the polarizer. Color coding and layout is analogous to the one in Fig. 5.10. *Left* The *blue regions* rotate with the reference frame but never show features of entanglement along the beam axis. *Right* The entanglement pattern of the sum over all tested reference frames reveals the novel feature of vector photons again, namely, that they are entangled and separable in polarization, dependent on the transverse region

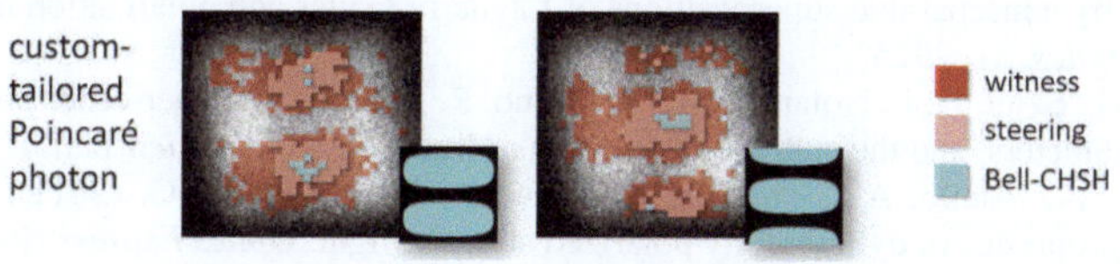

Fig. 5.13 Polarization entanglement pattern of custom-tailored Poincaré photons entangled with uniformly-polarized photons. Again, three different entanglement criteria were applied and the colors depict successfully tested regions (>100 photons contribute to every term of the criterion and the classical bound is exceeded by 3 standard deviations). *Left* All measurements were performed with H, V, D, and A polarized trigger photons. *Right* Measurements were conducted with D, A, R, and L trigger polarizations

that custom-tailored Poincaré photons are entangled everywhere but at the left and right edges of the square-shaped pattern. The same reasoning holds as before: At the left and right edges, the modal overlap is too small. Thus, the path inside the transfer setup, and thereby the polarization, is always known and no entanglement will be detected. Because this is only true at the left and right edge of the beam, custom-tailored Poincaré photons are entangled in the center. Therefore, they are also separable and entangled in the polarization DOF simultaneously.

The efficient generation and detection of the complex polarization pattern, as well as the surprising entanglement structures, were only possible due to the efficient way of creating and measuring the spatial properties of entangled pairs. Possible future investigations that take advantage of the presented results and further increase the complexity of the state will be briefly discussed in the next and last chapter.

References

1. Fickler, R., Lapkiewicz, R., Ramelow, S., & Zeilinger, A. (2013). Quantum entanglement of complex photon polarization patterns in vector beams. *Physical Review A, 89*(6), 060301(R).
2. Sondermann, M., et al. (2007). Design of a mode converter for efficient light-atom coupling in free space. *Applied Physics B, 89*, 489.
3. Andrews, D. L., & Babiker, M. (2012). *The angular momentum of light*. Cambridge: Cambridge University Press.
4. Machavariani, G., Lumer, Y., Moshe, I., Meir, A., & Jackel, S. (2008). Spatially-variable retardation plate for efficient generation of radially- and azimuthally-polarized beams. *Optics Communications, 281*, 732.
5. Tidwell, S. C., Ford, D. H., & Kimura, W. D. (1990). Generating radially polarized beams interferometrically. *Applied Optics, 29*, 2234.
6. Marrucci, L., Manzo, C., & Paparo, D. (2006). Optical spin-to-orbital angular momentum conversion in inhomogeneous anisotropic media. *Physical Review Letters, 96*, 163905.
7. Cardano, F., et al. (2012). Polarization pattern of vector vortex beams generated by q-plates with different topological charges. *Applied Optics, 51*, C1.
8. Zhan, Q. (2009). Cylindrical vector beams: from mathematical concepts to applications. *Advances in Optics and Photonics, 1*, 1.
9. Maurer, C., Jesacher, A., Fürhapter, S., Bernet, S., & Ritsch Marte, M. (2007). Tailoring of arbitrary optical vector beams. *New Journal of Physics, 9*, 78.
10. Galvez, E. J., Khadka, S., Schubert, W. H., & Nomoto, S. (2012). Poincaré-beam patterns produced by nonseparable superpositions of Laguerre-Gauss and polarization modes of light. *Applied Optics, 51*, 2925.
11. Milione, G., Sztul, H. I., Nolan, D. A., & Alfano, R. R. (2011). Higher-order poincaré sphere, stokes parameters, and the angular momentum of light. *Physical Review Letters, 107*, 053601.
12. Holleczek, A., Aiello, A., Gabriel, C., Marquardt, C., & Leuchs, G. (2011). Classical and quantum properties of cylindrically polarized states of light. *Optics Express, 19*, 9714.
13. Berry, M. V., & Hannay, J. H. (1977). Umbilic points on Gaussian random surfaces. *Journal of Physics A: Mathematical and General, 10*, 1809.
14. Leach, J., Dennis, M. R., Courtial, J., & Padgett, M. J. (2005). Vortex knots in light. *New Journal of Physics, 7*, 55.

Chapter 6
Conclusion and Outlook

6.1 Conclusion and Summarizing Remarks

In summary, this thesis presented a novel flexible way to generate and verify the entanglement of two photons in complex transverse spatial structures. It describes the results of three experiments in which a novel transfer setup was developed, a coincidence-imaging technique was established and both schemes were used to investigate novel properties of complex mode structures such as vector photons.

The rather simple but powerful idea was to start with high-quality polarization entanglement and transfer it in a second step to the transverse spatial degree of freedom. This transfer was done in a phase-stable interferometric scheme, which can be easily adjusted (by simply reprogramming the SLM) to generate any complex transverse spatial mode. In a first experiment, the transfer setup was used to generate photons that are entangled in very high orbital angular momenta. To verify the created entanglement, a novel measurement technique was developed, which takes advantage of the intensity structure and its connection to the phase of the state. In addition, those "slit-wheels" reveal a possible application of entangled photons with high OAM quanta, namely the enhancement in remote determination of angular positions.

In a second experiment, the investigation of entanglement in complex mode structures was extended by using an intensified CCD camera. One photon was measured in its polarization state while the other photon was sent through the transfer setup. The detection signal of the first photon was then used as a trigger for the intensified CCD camera, which in turn imaged (in coincidence) the spatial structure of the entangled partner photon. Because imaging with CCD cameras offers a high spatial resolution, it was possible to demonstrate the general flexibility of the transfer setup by creating entanglement in any spatial mode, even modes with highly complex structures. In addition, the efficiency of the whole setup was sufficiently high to record many single-photon events within a short time period. This made it possible to image the effect of entanglement in real-time.

R. Fickler, *Quantum Entanglement of Complex Structures of Photons*,
Springer Theses, DOI 10.1007/978-3-319-22231-8_6

In a last experiment, the transfer setup was adjusted to create more complex mode structures, i.e. vector photons with a transverse spatially varying polarization. The coincidence-imaging technique was extended to be polarization-sensitive, hence it could measure the transverse polarization pattern. As in the second experiment, the polarization measurement of one photon was used as a trigger for the ICCD camera, which recorded the transferred partner photon revealing its spatial structure. Again, the high spatial resolution of the camera as well as the high overall efficiency of the scheme made it possible to reconstruct complex polarization patterns from single-photon coincidence images. Additionally, three different entanglement tests were performed and their varying strengths have been visualized. Thereby an interesting novel feature of entangled vector photons have been found: depending on the transverse spatial position vector photons are both "entangled" and "not entangled" in polarization.

6.2 Outlook

Having established two novel experimental schemes of creating and imaging entanglement, it is interesting to consider possible future applications taking advantage of the developed techniques.

To further increase the complexity of the generated entangled state, it is worthwhile to consider more than two photons. Analogous to the bipartite experiment, polarization entanglement between three or four photons could be the starting point. Then, each of the photons might be transferred to the spatial degree of freedom by the method presented here. This would lead to multi-partite OAM entanglement that has not been observed to date. However, one important advantage of the transverse spatial degree of freedom over polarization, namely its potential for high-dimensional entanglement, cannot be addressed with such an experiment. The multi-partite entangled state would "only" be entangled in two-dimensional Hilbert spaces (since polarization entanglement would be the starting point again). In order to fully exploit the potential of high-dimensional spatial mode entanglement between more than two parties, other approaches have to be developed. Nevertheless, interesting experimental tests could already be envisaged, e.g. the swapping of entanglement between two different degrees of freedom. Here, the polarization-encoded photons of two hybrid-entangled pairs (as described in the earlier chapters of this thesis) could be projected into the Bell-state basis, which swaps the entanglement to the other two OAM-encoded photons (analog to the original idea of entanglement swapping described in [1–3]).

A second promising idea might be to combine the presented ideas with a long-distance experiment involving OAM states, similar of the recent experiment in [4]. These results, in combination with the efficiency of the transfer setup and the quality of OAM-encoded states, suggest that entangled OAM photons can be sent over a three-kilometer free space link within strong turbulent atmosphere. The presented schemes for detecting entanglement, either using the coincidence-imaging technique or masking the received photons (analog to the slit wheels), appear to be a realistic option, especially because it is not highly sensitive to turbulence induced beam wander.

Another interesting future experiment could further extend the limit of quantum entanglement with respect to high quantum numbers by following the idea introduced in Chap. 3. In this thesis, the entanglement of high OAM values has been explored up to values of $300\hbar$. However, this number was limited by the technical ability of the SLM to generate high orders of OAM containing modes. Novel technical developments, like the recently introduced spiral phase mirror by Prof. Lam and co-workers [5, 6], could replace the SLM and could be used to entangle even higher OAM quanta. We begun a collaboration with the group of Prof. Lam and tested spiral phase mirrors with a topological charge up to $2000\hbar$. First results concerning the entanglement of photons with such high quanta of OAM are very promising and will be investigated in the near future.

In general, it will be interesting to see which of the developed techniques will be used in future experiments to further investigate complex structures of photonic modes, thereby gaining a deeper understanding of quantum mechanics and the nature of light.

References

1. Bennett, C. H., et al. (1993). Teleporting an unknown quantum state via dual classical and Einstein-Podolsky-Rosen channels. *Physical Review Letters, 70*, 1895.
2. Żukowski, M., Zeilinger, A., Horne, M. A., & Ekert, A. K. (1993). "Event-ready-detectors" bell experiment via entanglement swapping. *Physical Review Letters, 71*, 4287.
3. Pan, J.-W., Bouwmeester, D., Weinfurter, H., & Zeilinger, A. (1998). Experimental entanglement swapping: Entangling photons that never interacted. *Physical Review Letters, 80*, 3891.
4. Krenn, M., Fickler, R., Fink, M., Handsteiner, J., Malik, M., Scheidl, T., Ursin, R., & Zeilinger, A. (2014). Communication with spatially modulated light through turbulent air across Vienna. New *Journal of Physics, 16*(11), 113028.
5. Campbell, G., Hage, B., Buchler, B., & Lam, P. K. (2012). Generation of high-order optical vortices using directly machined spiral phase mirrors. *Applied Optics, 51*, 873.
6. Shen, Y., et al. (2013). Generation and interferometric analysis of high charge optical vortices. *Journal of Optics, 15*, 044005.

Appendix A
Theoretical Formalism of the Slit-Wheel Measurement

The following theoretical description was developed by William N. Plick. It is given as an appendix to the thesis for the sake of completeness. First, the correlation as a function of angle due to OAM entanglement will be derived, followed by a theoretical model of the slit-wheel masks and the probability of the joint detection behind them.

As described in Chap. 3, an anti-correlated bi-photon state is generated. Each photon has a specific quanta l of OAM

$$|\Psi^{OAM}\rangle = \frac{1}{\sqrt{2}}\left(|+l\rangle\,|-l\rangle + e^{i\varphi}\,|-l\rangle\,|+l\rangle\right). \tag{A.1}$$

This state is maximally entangled in the OAM basis and spans a two-dimensional subspace of the full infinite-dimensional OAM Hilbert space. First, the angular correlation between the two photons a and b (labeled by the positions of the ket-vectors) will investigated.

A.1 Angle Correlation due to OAM Entanglement

The state from Eq. (A.1) can be rewritten in terms of Fock state excitations for both photons which leads to

$$|\Psi^{OAM}\rangle = \frac{1}{\sqrt{2}}\left(|1\rangle_{l,a}\,|0\rangle_{l,b}\,|0\rangle_{-l,a}\,|1\rangle_{-l,b} + |0\rangle_{l,a}\,|1\rangle_{l,b}\,|1\rangle_{-l,a}\,|0\rangle_{-l,b}\right), \tag{A.2}$$

where the subscripts indicate the OAM value ($\pm l$) and the spatial mode (a, b). It is useful to define the creation and annihilation operators for these modes. Starting with the annihilation operator as a function of angular position

$$\hat{a}_l(\gamma)\,|n\rangle_l = e^{il\gamma}\sqrt{n}\,|n-1\rangle_l\,, \tag{A.3}$$

R. Fickler, *Quantum Entanglement of Complex Structures of Photons*,
Springer Theses, DOI 10.1007/978-3-319-22231-8

where the phase term $l\gamma$ associated with the annihilated photon is given by the phase distribution of the LG modes and γ represents an angular position in the plane of absorption (detection). Because the photon may be absorbed by a detector, no matter its OAM, the annihilation operator for detection should be written as

$$\hat{a}(\gamma) = \frac{1}{\sqrt{2}}\left(\hat{a}_l(\gamma) + \hat{a}_{-l}(\gamma)\right). \tag{A.4}$$

This describes a photon being absorbed at a particular angular position γ. By requiring that $\hat{a}^\dagger\hat{a}$ be defined as the number operator and that $[\hat{a}, \hat{a}^\dagger] = 1$, the creation operator can be defined as

$$\hat{a}_l^\dagger(\gamma)\,|n\rangle_l = \sqrt{n+1}e^{-il\gamma}\,|n+1\rangle_l\,. \tag{A.5}$$

Similar creation and annihilation operators can be defined for the second photon $\hat{b}$ and $\hat{b}^\dagger$.

The coincident detection of both photons, as a function of angular positions γ_a (first photon) and γ_b (second photon), can be found as

$$\begin{aligned}\langle\hat{a}^\dagger\hat{a}\hat{b}^\dagger\hat{b}\rangle &= \langle\Psi^{OAM}|\left(\hat{a}_l^\dagger\hat{a}_l + \hat{a}_l^\dagger\hat{a}_{-l} + \hat{a}_{-l}^\dagger\hat{a}_l + \hat{a}_{-l}^\dagger\hat{a}_{-l}\right)\\ &\quad\times\left(\hat{b}_l^\dagger\hat{b}_l + \hat{b}_l^\dagger\hat{b}_{-l} + \hat{b}_{-l}^\dagger\hat{b}_l + \hat{b}_{-l}^\dagger\hat{b}_{-l}\right)|\Psi^{OAM}\rangle\,,\\ &= \cos^2 l(\gamma_a - \gamma_b).\end{aligned} \tag{A.6}$$

The single counts are constant up to a normalization.

As described intuitively in the main text (by assuming the slit-wheels to "project" onto a superposition state) an interference pattern with $2l$ fringes is displayed as a function of angular positions in the coincidence counts. Given this result, the measurement scheme of testing generated OAM state (A.1) by implementing two slit-wheels with the appropriate number of slits, is a natural choice. The bucket detectors behind the masks register the transmitted photons, hence they measure the photons to be at a specific angular position. If the coincidence patterns, recorded for different angular position of the two slit-wheels, matches the expected fringes, the generation of the target state can be concluded.

A.2 Slit-Wheel Measurement of OAM Entanglement

This procedure—though simple, intuitive, and inexpensive—has to be put into a mathematical framework for the slit-wheel as a quantum device. The slit-wheel mask is defined to act in the Hilbert space of angular spatial modes, i.e. modes of thin wedges of the space that the photons in the light beam can occupy. Thus, the projection operator for the first slit wheel may be written as

$$\hat{P}_1 = \sum_{n=0}^{2l-1} \int_{\frac{\pi n}{l}}^{\frac{\pi}{l}(n+W_s)} d\gamma_1 \left|\gamma_1\right\rangle \left\langle\gamma_1\right| , \tag{A.7}$$

where W_s is the slit width as a fraction of the distance to the next slit,[1] and the $|\gamma_a\rangle$'s are the angular position states. The mask for the second photon can be defined similarly by

$$\hat{P}_2 = \sum_{n=0}^{2l-1} \int_{\frac{\pi n}{l}+\Delta\gamma}^{\frac{\pi}{l}(n+W_s)+\Delta\gamma} d\gamma_2 \left|\gamma_2 - \frac{\pi}{2l}\right\rangle \left\langle\gamma_2 - \frac{\pi}{2l}\right| , \tag{A.8}$$

where, the shift in the projector states by $\pi/2l$ indicates that the wheel should be shifted one half-cycle. This relabeling leads to an anti-correlation of the statistics. Analog to the main text the factor $\Delta\gamma$ represents the relative difference in angle between the slit wheels.

To relate the angular position to LG modes, we write the LG states in terms of angular position states

$$|l\rangle = \mathcal{N} \int_0^{2\pi} d\gamma e^{il\gamma} \left|\gamma\right\rangle . \tag{A.9}$$

The phase factor originates from the transverse phase profile of the LG modes. The normalization factor $\mathcal{N}$ is found by requiring that $\langle l|l\rangle = 1$ resulting in $1/\sqrt{2\pi}$. The normalization of the projection operator is found to be unity by requiring that $\langle l|\, \hat{P}\, |l\rangle = W_s$.

The target state, Eq. (A.1), can then be rewritten as in the angular basis

$$\begin{aligned} |\Psi^{OAM}\rangle &= \frac{1}{\sqrt{2}} \left(|l,-l\rangle + |-l,l\rangle\right), \\ &= \frac{1}{2\pi\sqrt{2}} \left(\int_0^{2\pi} d\gamma_a e^{il\gamma_a} |\gamma_a\rangle \int_0^{2\pi} d\gamma_b e^{-il\gamma_b} |\gamma_b\rangle \right. \\ &\quad \left. + \int_0^{2\pi} d\gamma_{a'} e^{-il\gamma_{a'}} |\gamma_{a'}\rangle \int_0^{2\pi} d\gamma_{b'} e^{il\gamma_{b'}} |\gamma_{b'}\rangle \right) . \end{aligned} \tag{A.10}$$

[1] In the main text the inverse ratio, i.e. the ratio of distance between two slits and slit width, was used for better describing the effect of slit width to visibility. However, now it is more convenient to define the inverse ratio W_s.

Now, the joint probability of having a detection in both modes when the slits are turned to different angles can be calculated as

$$\begin{aligned} P_2(\gamma_a^+, \gamma_b^+) &= \langle \psi | \hat{P}_a \hat{P}_b | \psi \rangle, \\ &= \frac{1}{\pi^2} \left(\pi^2 W_s^2 - \cos(2l\Delta\gamma) \sin^2(\pi W_s) \right). \end{aligned} \tag{A.11}$$

The symbol $\gamma_{a/b}^+$ represents a click when the slit wheel is turned to position $\phi_{a/b}$. From this result, it can be seen, that the coincidences (joint detection) only depend on the relative angle difference $\Delta\gamma$ and not on the individual angular positions of each mask. Moreover, it can be seen that the finite width of the slit wheel only lowers the expected coincidences, thus it can be discarded to reduce the number of assumptions needed to prove the entanglement of the state.

Hence the slit-wheel masks installed in both beams can be used to test for entanglement of the two-dimensional entangled bi-photon state of high OAM.

Curriculum Vitae

Personal Information

Name	Robert Alexander Fickler
Date of Birth	4 January 1980
Place of Birth	Kempten (Allgäu), Germany

Professional Experience

Since 2015	Postdoctoral Fellow Department of Physics and Max Planck Centre for Extreme and Quantum Photonics, University of Ottawa, Ottawa, Canada
2014–2015	Postdoctoral Fellow Faculty of Physics, University of Vienna, Vienna, Austria IQOQI Vienna, Austrian Academy of Sciences, Vienna, Austria
2009–2014	Research assistant at the Faculty of Physics, University of Vienna, Vienna, Austria
2012	Research assistant at the School of Mathematics and Physics, University of Queensland, Brisbane, Australia
2008–2009	Research assistant at the Institute for Quantum Information, University of Ulm, Ulm, Germany
2006–2009	Tutor for undergraduate and graduate students, University of Ulm, Ulm, Germany
2000–2001	Alternative civilian service Municipality of Kempten, Kempten, Germany

R. Fickler, *Quantum Entanglement of Complex Structures of Photons*,
Springer Theses, DOI 10.1007/978-3-319-22231-8

1996–1999 Apprenticeship as an electronics engineer,
Allgäuer Überlandwerk, Kempten, Germany

Education

9/2014 PhD (Doktorat) degree in physics,
Faculty of Physics, University of Vienna, Vienna, Austria
Prae Doc
CoQuS-Fellow
(Vienna Doctoral Program for Complex Quantum Systems),
4/2009 Master degree (Diplom Physik) in physics,
University of Ulm, Ulm, Germany
5/2006 Bachelor degree in philosophy,
University of Ulm, Ulm, Germany
6/2002 A-levels exam (Abitur),
Technical college for further education, Kempten, Germany
7/1996 O-levels exam (Mittlere Reife),
Secondary school, Kempten, Germany

Awards and Distinctions

- **Young Scientist Prize 2015** of the IUPAP C17 (Laser Physics and Photonics) for outstanding contributions of a fundamental nature
- **Doc.Award 2014** for outstanding doctoral theses (University of Vienna and City of Vienna)
- **Best Paper Award 2013** of the City of Vienna for "Quantum entanglement of high angular momenta"
- "Quantum entanglement of high angular moment" named as one of the **Top 10 Breakthroughs of the Year 2012** (IOP's Physics World)

Publications

2015

M. Krenn, J. Handsteiner, M. Fink, R. Fickler, A. Zeilinger "Twisted photon entanglement through turbulent air across Vienna", arXiv preprint arXiv:1507.06551 (2015)

W. N. Plick, R. Fickler, R. Lapkiewicz, S. Ramelow "Violation of an extended Wigner inequality with high-angular-momentum states" Physical Review A **91**, 022124 (2015)

2014

M. Krenn, R. Fickler, M. Fink, J. Handsteiner, M. Malik, T. Scheidl, R. Ursin, A. Zeilinger "Communication with spatially modulated light through turbulent air across Vienna" New Journal of Physics **16**, 113028 (2014)

R. Fickler, R. Lapkiewicz, M. Huber, M.P.J. Lavery, M.J. Padgett, A. Zeilinger "Interface between path and orbital angular momentum entanglement for high-dimensional photonic quantum information" Nature Communications **5**, 4502 (2014)

M. Krenn, M. Huber, R. Fickler, R. Lapkiewicz, S. Ramelow, A. Zeilinger "Generation and confirmation of a (100 × 100)-dimensional entangled quantum system" Proceedings of the National Academy of Sciences **111**, 6243–6247 (2014)

R. Fickler, R. Lapkiewicz, S. Ramelow, A. Zeilinger "Quantum entanglement of complex photon polarization patterns in vector beams" Physical Review A **89**, 060301(R) (2014)

2013

R. Fickler, M. Krenn, R. Lapkiewicz, S. Ramelow, A. Zeilinger "Real-time imaging of quantum entanglement" Scientific Reports **3**, 1914 (2013)

S. Rahimi-Keshari, M. A. Broome, R. Fickler, A. Fedrizzi, T. C. Ralph, A. G. White "Direct characterization of linear-optical networks" Optics Express **21** (11), 13450–13458 (2013)

W. N. Plick, M. Krenn, R. Fickler, S. Ramelow, A. Zeilinger "Quantum orbital angular momentum of elliptically symmetric light" Phys. Rev. A **87**, 033806 (2013)

M. Krenn, R. Fickler, M. Huber, R. Lapkiewicz, W. N. Plick, S. Ramelow, A. Zeilinger "Entangled singularity patterns of photons in Ince-Gauss modes" Physical Review A **87**, 012326 (2013)

2012

R. Fickler, R. Lapkiewicz, W. N. Plick, M. Krenn, C. Schaeff, S. Ramelow, A. Zeilinger "Quantum entanglement of high angular momenta" Science **338**, 640–643 (2012)

C. Schaeff, R. Polster, R. Lapkiewicz, R. Fickler, S. Ramelow, A. Zeilinger "Scalable fiber integrated source for higher-dimensional path-entangled photonic quNits" Optics Express **20** (15), 16145–16153 (2012)

2010

W. Schnitzler, G. Jacob, R. Fickler, F. Schmidt-Kaler, K. Singer "Focusing a deterministic single-ion beam" New Journal of Physics **12** (6), 065023 (2010)

2009

R. Fickler, W. Schnitzler, N. M. Linke, F. Schmidt-Kaler, K. Singer "Optimised focusing ion optics for an ultracold deterministic single ion source targeting nm resolution" Journal of Modern Optics **56** (18-19), 2061–2075 (2009)

W. Schnitzler, N. M. Linke, R. Fickler, J. Meijer, F. Schmidt-Kaler, K. Singer "Deterministic ultracold ion source targeting the Heisenberg limit" Physical Review Letters **102** (7), 70501 (2009)